"Intriguing . . . challenges commonsense notions of what time and space really are. After reading about these concepts, you may wonder just what reality really is." —*Booklist*

Today a trip to the moon takes about three days, a journey to Mars takes several months, and a voyage to Earth's nearest star, Alpha Centauri, would take almost a million years. Exploration of the universe is hampered both by the vastness of space and the slowness of existing means of travel. But if wormholes—"cosmic gateways" between regions of the universe billions of miles apart—prove to be a reality, the Cooks and Magellans of the future will be able to span unimaginable distances relatively quickly and explore realms we can now only dream of. In this absorbing book, Paul Halpern takes readers on a journey to the frontiers of the scientific imagination, as he explains the fascinating implications of wormhole theory. In the future we may be able to surprise H. G. Wells with a visit from an actual time machine—but for the moment, a good book is the best way to soar across the cosmos.

"Colorful . . . inspiring . . . Halpern tackles the No. 1 recreational problem of theoretical physics: actual applications of special relativity to create the possibility of spacetime travel." —*Publishers Weekly*

PAUL HALPERN holds a Ph.D. in theoretical physics from the State University of New York at Stony Brook. Currently an associate professor of mathematics and physics at the Philadelphia College of Pharmacy and Science, he has written numerous articles on chaos theory, cosmology, and the history of science. He is also the author of *Time Journeys: A Search for Cosmic Destiny and Meaning*. He lives in Philadelphia.

COSMIC WORMHOLES

THE SEARCH FOR INTERSTELLAR SHORTCUTS

PAUL HALPERN

A PLUME BOOK

PLUME
Published by the Penguin Group
Penguin Books USA Inc., 375 Hudson Street, New York, New York 10014, U.S.A.
Penguin Books Ltd, 27 Wrights Lane, London W8 5TZ, England
Penguin Books Australia Ltd, Ringwood, Victoria, Australia
Penguin Books Canada Ltd, 10 Alcorn Avenue, Toronto, Ontario, Canada M4V 3B2
Penguin Books (N.Z.) Ltd, 182–190 Wairau Road, Auckland 10, New Zealand

Penguin Books Ltd, Registered Offices: Harmondsworth, Middlesex, England

Published by Plume, an imprint of Dutton Signet,
a division of Penguin Books USA Inc. Previously published in a Dutton edition.

First Plume Printing, October, 1993
10 9 8 7 6 5 4 3 2 1

Ⓟ REGISTERED TRADEMARK—MARCA REGISTRADA

LIBRARY OF CONGRESS CATALOGING-IN-PUBLICATION DATA
Halpern, Paul, 1961–
    Cosmic wormholes : the search for interstellar shortcuts / Paul
Halpern.
        p.   cm.
    ISBN 0-452-27029-4
    1. Astrophysics.   2. Space flight.   I. Title.
QB461.H24   1993
523.01—dc20   93–17393
                                                     CIP

Printed in the United States of America
Original hardcover design by Leonard Telesca

BOOKS ARE AVAILABLE AT QUANTITY DISCOUNTS WHEN USED TO PROMOTE PRODUCTS OR
SERVICES. FOR INFORMATION PLEASE WRITE TO PREMIUM MARKETING DIVISION, PENGUIN
BOOKS USA INC., 375 HUDSON STREET, NEW YORK, NEW YORK 10014.

To Stanley and Bernice, my parents

ACKNOWLEDGMENTS

I'd like to thank Graham P. Collins of *Physics Today* for many interesting discussions about traversable wormholes and for introducing me to some of the most recent discoveries in this field. In addition, many thanks to Fred Schuepfer for supplying me with the "physicist in the chicken factory" joke (apparently one that is in vogue among civil engineers) and to Michael Erlich for numerous philosophical discussions. Thanks also to Bernard Brunner, Charles Gibley, Dolores Orensky, David Kerrick, William Walker, Bill Reinsmith, and other members of the faculty and staff of the Philadelphia College of Pharmacy and Science for their strong support, as well as to Rachel Klayman for her editorial assistance and numerous helpful suggestions. Special thanks to my agent, John Ware, a man of exceptionally high principle, talent, and insight. Finally, I gratefully acknowledge the considerable moral support of my family and friends, including Richard, Alan, Kenneth, Bernice, Stanley, and Esther Halpern, as well as Tamara White, Kris Olson, Dubravko Klabucar, Boris Briker, Alex Schwartz, Evan Thomas, Carolyn Brodbeck, Bill Schwartz, Ben Genaro, Tom Caltagirone, Robert Clark, and Bozena Strak.

CONTENTS

COSMIC
WORMHOLES

INTRODUCTION

In the past few decades, humankind has embarked on an exploration of the vast reaches of space in search of possible new forms of life. Yet it has become increasingly clear that these efforts are severely impeded by two formidable barriers: the enormity of space and the sluggish pace of space travel. Given the current speed of spaceships, a trip to Mars, for instance, would take several months to complete, and a journey to the nearest star, Alpha Centauri, would take hundreds of thousands of years. Voyages to planets in other solar systems, including those believed to be capable of sustaining life, could take literally millions of years. Clearly, present technology simply cannot accommodate man's aspirations to soar through the cosmos with ease and sample the periphery of his world.

Is there any hope for interstellar contact in the future? Although there is a good chance that new means of rapid transport utilizing far more powerful sources of energy than used at present will someday be developed, it is nevertheless unlikely that speeds will be improved significantly. Furthermore, Einstein's time-tested theory of special relativity places a strict upper limit—the speed of light—on the rate of such journeys. According to this theory, light propagation, in the form of visible rays or invisible radio transmissions, is necessarily faster than the velocity of matter. Hence a space voyage to another planet must take longer than the passage of light from earth to that planet. This span of time, in most cases, would be tens or hundreds of years. Therefore, conventional rocket travel of any degree of sophistication would be highly impractical for long journeys.

But physicists are a resourceful lot and are not inclined to give up easily on a problem such as space travel. A few years ago, while working on the novel *Contact*, Carl Sagan pondered the question of whether or not members of alien civilizations could travel to the earth, given the known constraints of astrophysics. Understandably, Sagan wanted to put a bit of realism into his science fiction epic by describing such a voyage in accurate physical detail. Unable to find a satisfactory answer to this question himself, he called up one of his best friends, the physicist Kip Thorne of the California Institute of Technology.

Kip Thorne is a cosmologist known for his creativity and his unique ability to explain complex physical phenomena in a simple manner. Being also a man who enjoys a challenge, he attacked Sagan's question with great vigor, calling in his graduate student Michael Morris for extra help. Together, Morris and Thorne worked hard on a theoretical scheme for feasible interstellar travel.

The theory they devised is truly revolutionary in its nature. They elegantly proved that there exist theoretical celestial objects, called *wormholes*, which bridge separate parts of the universe in such a manner that direct travel is possible between any two linked sections. In other words, in Morris's and Thorne's view, a cosmic gateway could be created between regions of the universe billions of billions of miles apart that would allow nearly instantaneous transport between these regions. Thus the mind-boggling distances normally involved in transgalactic travel could be significantly reduced.

Since the publication of the theory of traversable wormholes there has been a massive amount of research in this field. Morris and Thorne's extraordinary concept has led to an unprecedented surge of interest in the idea of instantaneous transport. For the very first time, physicists can almost taste actuality in what has been an extremely speculative, if highly engaging, field. Only now can we realistically imagine the rapid transport of passengers and materials across the universe in networks of transgalactic commerce. Truly, the last few years have produced a significant theoretical breakthrough in astrophysics and a source of inspiration for our ultimate plans to explore space.

Most recently, the concept of wormholes has entered the popular culture as well. The television series *Star Trek: The Next Generation* has prominently featured the use of wormholes as a means for transgalactic passage. If, in this case, science fiction forecasts reality, as it often does with such remarkable precision, we can expect some wonderfully exciting times ahead.

In spite of its striking resemblance to the fantastic schemes of science fiction, the notion of wormholes is based on years of careful scientific preparation and inquiry. Although the traversable wormhole model in its current form is only three years old, the concept of using relativity theory to design space tunnels is considerably older. Scientists have long known that one of the predictions of Einstein's model is that extremely massive stars would collapse eventually to become peculiar objects called black holes. These mysterious celestial bodies, because of their small size and enormous gravitation, draw in all surrounding materials on a one-way course toward their centers. Even light does not escape their vicinity—hence, the name *black hole*.

Remarkably, one prediction of black hole theory is that these objects, if spinning, are connected in pairs by tunnels. In other words, according to the theory, each rotating black hole has a companion in another part of the universe (or in a different universe altogether). For a brief period of time, it was thought that vessels could travel through these tunnels and emerge thereby in another region of the cosmos. However, because of the enormous gravitational tidal forces present en route and the inability of vessels to escape them intact, it has since been concluded that attempting to travel between black holes would be a fatal blunder.

One attempt to remedy this theoretical drawback involved objects known as *white holes* (negative black holes). Just as the former are perfect absorbers of matter and energy, the latter are perfect emitters—"cosmic gushers." Mathematically it was found that one could theoretically travel into a black hole, pass through a connecting tunnel, and emerge intact from a white hole in another part of the universe (assuming that one could resist enormous gravitational tidal forces). Unfortunately, though, these white hole gateways would be unstable and almost immediately decay, rendering passage through them virtually impossible. Furthermore, it's unclear whether these theoretical constructs exist at all in nature.

Wormholes, of the sort envisioned by Morris and Thorne, have almost none of these drawbacks, since they are models designed specifically to provide safe and permanent passage. Noting that Einstein's theory of relativity allows for a tremendously wide range of possible celestial objects, Morris and Throne have selected designs that are, in theory, navigable by space vessels. Theoretically, traversable wormholes could be constructed today, given the right amount of matter in the proper configuration. Practically, however, we'll need to wait until our knowledge of space material is more exact. Thus, this newest theory, although far from being realized in practice, is indeed a dramatic leap forward in our understanding of astrophysics and represents a substantial advance beyond the older black hole model.

Not only could these wormholes be used for space flight: the theory predicts that they could be used for time travel as well. Morris and Thorne, along with the theorist Ulvi Yurtsever, have, for the first time in modern physics, designed a plausible theoretical model of a time machine. Specifically, they have proved that a traversable wormhole, if properly constructed, would grant an explorer free passage in time.

The implications are staggering. Like the time traveler in the famous H. G. Wells story, an intrepid voyager could enter a specially designed wormhole and emerge at any date in history. An unexpected offshoot of the work of the Caltech team, time machine construction is an intriguing theoretical endeavor.

The mere prospect of using wormholes for time travel creates some fascinating dilemmas. What would happen if an intrepid time traveler were to use one of these objects to go back in time and accidentally kill his own great-grand-mother? Would losing one of his ancestors cause him to vanish from the universe? If so, how could the murder have been committed? This rather morbid puzzle, known to physicists as the *time travel paradox*, raises interesting philosophical issues, many of which will be explored in this book.

Woody Allen, in his book *Without Feathers*, describes the opening of a new "express route" to a distant inhabited planet called Quelm. This "shortcut" slashes *two full hours* off the usual *6-million-year* journey from earth to the planet. I sincerely hope that the spatial shortcuts produced by traversable wormholes would save us at least as much time as the "express route" that Allen depicts.

Space and time travel are fascinating topics with almost universal appeal. I've been interested in these themes since childhood, when I read *The Time Machine* and *The First Men in the Moon*, both penned by the supreme master of science fiction, H. G. Wells. And since it is unlikely that wormhole technologies will be fully developed during our own lifetimes, I urge readers to obtain one of the inspiring epics by the likes of Wells, Verne, Bradbury, and company and to savor a clanky trip through the universe aboard a space kettle or such. Surely, at least until interstellar gateways are developed, a good book is the quickest way to soar across the cosmos.

Paul Halpern
Philadelphia

CHAPTER 1

LIFT-OFF

*A*loneness

Imagine that you are condemned to spend your life in a small town in the middle of a vast desert. This town is isolated in the extreme: there has been absolutely no contact with any other villages for centuries because the nearest are over a thousand miles away. Still, there are plenty of resources in your community—lots of food and water, several interesting places to go for entertainment, hundreds of people with whom to talk. There is never a shortage of activities and never a want of supplies. It is clear that the town could sustain itself for decades to come and that you could spend the rest of your life there in peace and contentment.

Sometimes, however, in the middle of the night you are struck by a terrifying notion. You realize that you may never know a single thing about the people and events in other communities; the vast world around you may be forever beyond your grasp. You imagine universes of thought and action that are spinning around the singleness of your tiny enclave in a silent, mysterious, and elusive dance. At these times you feel painfully alone and lost in the enormity of the unreachable space beyond your village.

When you wake up after these turbulent nights, a strong wanderlust descends upon you. Hastily, you set out in the desert, always proceeding at random, given your total lack of knowledge of any place out there. You walk for miles and miles, for days and days. Finally, you turn back; you never continue beyond a few weeks. It's hopeless, you think. The distances are much too great; you'll literally never get anywhere. You retreat to your enclave and try for a while to forget about the enormity of the world and the goings-on in the countless other towns.

One day, however, you and the other people of the town are startled by a strange noise. Running outside toward the sound you discover that an enormous 747 jet airplane has landed in the desert near the community. Because you have never seen an airplane before, you are amazed to find out that it offers the very thing you've been after: the means to travel to other towns. An airplane, you are told, could reduce to hours the desert journeys that, as you've rightly suspected, would take years on foot. Finally your isolation will be broken and you'll be able to explore the world around you. At last you are no longer alone in the universe.

Entering the plane, you bid one last goodbye to the town. The plane takes off.

Reaching Out

We all dwell on a vastly isolated and tiny earth, an oasis of humanity in an unknown desert of enormous proportions. For all we know, we could be the only living creatures in the cosmos; the universe could just as well be teeming with life. We are condemned, at least for now, to have little knowledge of the world beyond this fragile island.

When will our rescue plane arrive? When will we be liberated from this lonely planet? Rather than waiting for the slim possibility that alien life-forms will reach out and contact us, we had best develop means of our own to explore the far reaches of the cosmos. Yet it is not entirely clear how to undertake this great task.

The easiest regions of outer space to explore are the earth's suburbs, namely the other planets of the solar system. Unfortunately, these "nearby" worlds present us with empty wastelands. In spite of numerous unmanned missions to other planets, especially to Mars and Venus, no evidence for extraterrestrial life has been uncovered. Almost certainly, the prerequisites for complex life are not found on the earth's neighboring worlds. This is especially likely because of the special factors that must be present for intelligent life to exist and develop: just the right type of atmosphere, the correct distance from the sun, and the proper amount of surface water. Although scientists hoped for years that Mars would contain life, the space probes sent there during the seventies found absolutely no signs of living creatures or even fossils. The *Voyager* missions have shown us the extremely hostile atmospheric conditions on the other planets and moons. Therefore, although we still cannot rule out completely the existence of life in this region, we can consider it very unlikely.

Over the coming decades, we shall understand the solar system better and better. Unmanned space probes will continue to be sent out to most of the planets and moons, and these will be followed by manned missions in all likelihood. Sadly, though, it is probably the case that the solar system, which can be explored in relatively short periods relative to the rest of space, is not the most fertile ground for fundamental research into the nature of living organisms. To find life beyond our own we must go elsewhere, to other stellar systems, and this would take a vast amount of time at present speeds.

Given our current capabilities we can reach the moon, a distance of roughly a quarter of a million miles away, in about three days. This means that our rocket ships can travel at a rate of 25 million miles per year. Although this might seem like a phenomenal speed, at this rate it would take a million years to reach the nearest star, Alpha Centauri, which is almost 25 *trillion* miles from the earth. Thus the time scale difference between a lunar voyage and even the simplest interstellar journey is analogous to the distinction between the amount of time needed for a boy to build a model airplane and that required for humankind to evolve on earth!

Moreover, to journey beyond "neighbors," such as Alpha Centauri, to stars that are even farther away would require durations of time that are even longer. Our galaxy, the Milky

Way, has a diameter of roughly 1 quintillion (1 billion billion) miles and would take almost a trillion years to circumnavigate at present speeds. Traveling to other galaxies would take at least one hundred times longer. These time periods are so long that they surpass even the current age of the physical universe.

It's clear that increasing the rate of space flight tenfold or a hundredfold would not significantly alter the present impracticality of interstellar space travel. After all, who cares whether it takes 1 million years or merely 10,000 years to reach another planet if all of the passengers on the ship would be long dead? Once one is dealing with numbers that are longer than a human lifetime, space travel of the conventional sort becomes unrealistic.

Numerous ways have been suggested to get around the problem of the enormous distances involved in space flight. Some scientists have suggested that interstellar flights could involve generations of families situated in an "ark." This massive structure would contain dormitories, kitchens, recreational facilities, nurseries, and storage compartments. Ideally it would be a miniature earth, with greenhouses and farms producing the food, energy, and oxygen needed for life. All materials would be recycled as much as possible, since waste products would create a tremendous burden for such a voyage. The plants in the greenhouses would absorb the extra carbon dioxide that would accumulate on the vessel.

Because in free space all objects are weightless, an artificial sort of gravity would be necessary to maintain the muscular fitness of the passengers. Otherwise, it is likely that their muscles would atrophy because of disuse. This synthetic gravity could be produced by causing the vehicle to rotate. The rotation would generate the effect of a centrifuge and cause objects to be pulled toward the walls of the ark, simulating the effects of the earth's gravitational pull on terrestrial bodies.

Thousands of children would be born aboard the vessel, children who never would have experienced the green grass or blue skies of their ancestors' planet. It is quite possible that these children would be well adjusted to life in space and might even be more comfortable there than on earth. In time, after a lengthy, multigenerational journey, they would of course grow to be the leaders of the expedition.

Finally, the space ark would reach some distant planet orbiting some alien star, where all of the travel-weary passengers would disembark and begin new lives as space pioneers. This new world would in time become as familiar to them as was the earth to their long gone terrestrial ancestors.

Could a space ark sustain life for hundreds of years? Could volunteers be found for such a demanding mission? Clearly there are serious questions about the feasibility of developing such a program. It would certainly take a hardy and adventurous lot to choose to spend the rest of their lives in space. Any sort of environmental problem would result in the destabilization of the ark's delicate balance, causing illness and perhaps even death of all the passengers. Even if the mission were completely safe, there might still be substantial difficulties for the space pilgrims. Because they would be cut off completely from their home planet, it is likely that lack of variety, restlessness, and boredom would lead to the demoralization of the crew. Thus the organization of such a risky undertaking would be highly problematic at best.

A much more humane alternative to the space ark proposal relies on the process of *cryogenics*, or deep freezing. Using this procedure, astronauts would be placed in a state of suspended animation by drastically lowering the temperature of their bodies. Thus, all of their bodily systems would function at a greatly reduced rate. This would enable the astronauts to sleep through centuries of space flight needing little food and oxygen and simultaneously foregoing the boredom associated with such a long mission. On arriving in the vicinity of their destination, the passengers would be awakened by timers and their bodily processes would be restored to full levels of activity. The astronauts would emerge from their slumber refreshed and ready to explore the new worlds.

Unfortunately, cryogenics is still in its infancy. Nobody understands the nature of the unconscious state well enough to risk the possible consequences, such as brain damage, that such long periods of low temperature sleep might induce. So far this sort of freezing has been taken advantage of only by relatively wealthy people who have asked to be frozen after death in the hope that they might be revived in the far future. (A pervasive but false rumor is that Walt Disney was frozen for this reason.) Thousands have undergone this process post mortem and may in fact someday be resuscitated by an advanced civilization with superior medical techniques. However, at the present time the revival phase of cryogenics is confined to the realm of science fiction.

A million years is an unimaginably long time. Yet, as we have seen, conventional space travel requires just such a period of confinement aboard a finite vessel. We hope that as technology grows more and more sophisticated, advanced methods for preserving human life for such a prolonged period will be developed. Many scientific advances occur at an extraordinarily rapid pace. More than a century ago nuclear power was unimaginable; now it's ubiquitous in many parts of the developed world. Perhaps even as early as the next century, deep freezing or artificial ecosystems will allow the Cooks and Magellans of the future to explore new realms beyond our current reach.

Soaring through Space, Touring through Time

In 1968 and 1969, two remarkable projects enabled the public to imagine soaring above the clouds and into the darkness of space in ways they could have never envisioned before. One of these actually took place; the other was fictional. They both fueled further the burning desire to explore the cosmos. Millions of people around the planet, safely situated in either their living rooms or in movie theaters, tuned in to spectacular images of men conquering the vast divides between earth and the nether reaches of creation. With these visual marvels they felt like children gazing for the first time at a wondrous fireworks display.

All eyes were on Neil Armstrong as he stepped for the first time onto alien soil. Nobody knew with absolute certainty whether or not Armstrong and the others would sink into the moon's dust like ballast in quicksand; no one knew whether they would make it back to earth alive. Fortunately, as we know, the mission was a resounding success: the moon's surface turned out to be sufficiently solid to support the crew and the *Apollo 11* astronauts returned safely back to their home planet.

The mission to the moon was the culmination of a decade-long program instituted by President John F. Kennedy to stake out a new frontier in space. In order to achieve this goal, the United States spent millions of dollars in an intense effort, calling on some of the best scientists and technicians available at the time. The result that summer day was a feat that truly amazed the world.

Unfortunately, the space program of the 1970s and 1980s failed to live up to the expectations of the 1960s. Both public awe and support lessened, and the next major goal of the space program, a manned mission to Mars, was indefinitely postponed. In an age of limited resources, the staggering costs of manned space flight were increasingly hard to justify. Numerous unmanned missions to other parts of the solar system were launched and the shuttle program was founded, but somehow none of these ventures spawned great interest. Nothing, it seemed, could generate more excitement than having three young men walk on another world. Except for the relatively simple case of the moon, however, this was easier said than done—voyages to other planets required too much time and money.

Because of these factors, cinematic depictions of space flight far outpaced factual developments; obviously it was far easier to produce a film about interplanetary travel than to design an actual voyage. This was especially true of the year 1969. As the *Apollo* astronauts were heading out on their flight to the moon, earth dwellers were debating the meaning of a remarkable new film, *2001: A Space Odyssey*, the most fully fleshed-out account of space travel in the history of cinema. *2001*, written by Arthur C. Clarke and directed by Stanley Kubrick, captured the imagination of a public eager to learn about space travel of a more adventurous sort than the lunar mission. Millions of filmgoers flocked to this innovative motion picture and let their imaginations soar through space and tour through time, amazed and baffled beyond belief.

The film begins with an engaging re-creation of primitive man's first attempts to use tools. Suddenly, the story leaps into the future (the year 2001), where computers and space cruisers have replaced sticks and sharp-edged rocks as the instruments of human ingenuity. A lunar colony has discovered a strange signal being sent from one of Jupiter's moons. Scientists hastily organize a space mission to Jupiter to investigate this odd phenomenon.

The spaceship *Discovery*, assigned to complete this mission, consists of two parts: one static, the other revolving. The revolving section is designed to produce a centrifuge effect to simulate gravity. Five passengers operate the ship; three are frozen in devices called *hibernaculums*, while the other two are up and about. The cryonically suspended passengers are to be awakened when the ship nears Jupiter; until then their deep freeze helps to save food expenses and to solve the problem of boredom.

Besides the five human crew members there is a sixth participant: the computer assigned to monitor the journey. The computer is programmed to know exactly when to wake up the hibernating passengers and to keep the other two people healthy and entertained. In addition, it controls all of the communications to and from the ship.

We can see from this operation how cryogenics could indeed help reduce the problems associated with journeys within the solar system, but also how it might not suffice for longer trips through the cosmos. The hibernation works reasonably well on the *Discovery* mission, which takes several years; however, as we have seen, interstellar voyages would take a million times longer. Certainly, the *Discovery* ship is hardly equipped to go anywhere beyond Jupiter. Considering the costs involved in an interplanetary mission, it is truly sobering to ponder the much greater risks and expenses that would be associated with trips outside the solar system.

In fact, the leader of the *Discovery* mission, David Bowman, does manage to travel beyond the Jupiter target to distant stars. On one of Jupiter's moons, Bowman discovers an enormous rectangular monolith which serves, strangely enough, as a space portal leading to a remote part of the universe. Bowman enters the massive edifice in a space pod (small spacecraft) after his ship has become disabled near Jupiter. He finds to his bewilderment that its inside is hollow and appears full of stars. Some sort of current drags him through the monolith, and he arrives in a region where time and space have become inverted: that is, he can move rapidly through the latter but not at all through the former. His watch first slows down, then stops, as he continues to blast forward through this strange world. All the while, he is surrounded by eerie colors, bathing him in an otherworldly luminescence. Suddenly he finds himself in a vast "switching room," walls and floors lined with millions of copies of the bizarre monolith. Each of these duplicates serves as a passageway to another part of the cosmos. His space pod is drawn toward one of these replicas and he realizes that he is about to enter another portal. Passing through this cosmic turnstile, he returns to normal space once more. His watch starts up again—an indication that he has returned to normal time as well. But something is wrong: the stars look odd. None of the familiar constellations is in the sky. Even the cloudy tapestry of the Milky Way is absent from the heavens. Bowman realizes that he must indeed be a long, long way from home.

What has happened is that Bowman has passed through the Star Gate, a celestial clearinghouse for transgalactic passage. According to Clarke's story, this remarkable structure was built by a remote and highly intelligent alien civilization in order to expedite interstellar journeys. Bowman has literally traveled billions and billions of miles in a few minutes of his own clock time. The aliens have cleverly concocted a method by which their distant planet can be reached by earthlings in a relatively brief period. And the aliens' beacon has summoned Bowman to his final destiny (made clearer in the sequels).

Leaving the movie theater after this final scene, most viewers (including this one) were absolutely perplexed. The entire sequence involving the passage through the monolith and the arrival in the alien world seemed like a bizarre multicolored phantasm, too complex to be sorted out satisfactorily. Many of those who saw the film dismissed this part as being an inconsequential footnote to an otherwise visionary project. But the few individuals with the imagination and foresight to understand the significance of Bowman's incredible journey through the portal were astounded by its real-life implications.

Imagine what would happen if a Star Gate were to exist. As a cosmic shortcut it would eliminate the need for lengthy passage and costly cryonic suspension within a space ark. Instead of a 1-million-year ordeal, travel through the universe would become almost as easy as a ride on the subway; one would simply pass through the interstellar turnstile, hop on, and go. With duplicate Star Gates in other solar systems, passengers could select their stops with discrimination and could explore the cosmos at their leisure. Our world would be integrally connected with others in a vast transportation network.

It is clear that of the two methods for space travel presented in *2001: A Space Odyssey*, slow-moving rocket-powered vessels carrying hibernating crew members and rapid-functioning interstellar portals transporting alert passengers, the latter would be far preferable. Other films made in recent years paint similar pictures: far-flung space journeys involving prolonged intervals of confinement, costing voyagers years of their lives. Consider, for example, the situation presented in another late 1960s science fiction movie, *Planet of the Apes*. In this film, a group of astronauts sets out on a journey to a distant star. Because of the extreme length of their travels in space, they return to earth hundreds of years after they leave. They find on arrival a much transformed society (governed by gorillas, chimpanzees, and orangutans). By making this space voyage they have, in effect, traveled through time, and they now face the sad result: the prospect of being stranded in the future, with all of their families long dead and no hope of ever seeing them again. The obvious question is, Was the journey worth it for the astronauts, just to explore a new region of space? Certainly not, in this case.

Any sort of lengthy space voyage would involve this sort

of moral dilemma. How could we ask our space pioneers to leave behind their loved ones, perhaps forever, to travel hundreds, thousands, or even millions of years into the future? Who would be audacious enough to demand this ultimate sacrifice: the severing of all connections to family, friends, and earth itself? Clearly, a much more compassionate alternative must be sought. That alternative is the Star Gate.

Beyond the Star Gate

Somewhere, far beyond the wispy clouds that cover the surface of the earth, farther even than the silent orbit of Pluto, perhaps even as distant as the center of the galaxy, there lies an unknown world: the closest inhabitable planet to earth. This world may be teeming with life, or it may be dead; it may contain primitive cultures that remind us of our past or advanced civilizations that suggest our own likely future. It could be covered with desert or ocean, forests or fields, bizarre new species of plants and animals, or creatures similar to terrestrial flora and fauna. Vast treasure houses of minerals might be buried beneath its surface, with incredible fossils hidden within its caves. New and powerful medicines might be found within the genetic material of its life-forms. Without a doubt, the discovery of this new planet would herald revolutions in biology, geology, anthropology, and physics, possibly spawning the emergence of a whole new order in the sciences. To find it would be one of the pivotal events in the history of humanity, rivaling even the discovery of fire and the invention of the wheel.

So how do we know that another such inhabitable planet exists in the universe? Of course, we don't know it with absolute certainty; we can merely venture a guess. Tossing a coin a thousand times, chances are that the coin will land on heads at least once; we'd be absolutely shocked if the results were 100 percent tails, stunned by an apparent violation of the laws of probability. But we would realize that these results, however unlikely, are within the realm of statistical possibility. Similarly, there is a *chance* that no other inhabitable planets exist in the cosmos, but the odds of this are far lower even than those of a thousand-tail coin toss. In fact, it can be shown that the odds for the existence of extraterrestrial life in the universe are fairly high.

Consider that there are at least a few hundred billion stars in the Milky Way galaxy. Current estimates, based on careful star counts, place this figure at 4×10^{11} suns (that's 4 followed by eleven 0's; in the exponential notation that I'll be using throughout this book the superscript refers to the number of zeroes following the whole number). Conservatively, at least one out of ten of these stars is orbited by planets; that is, forms a solar system. This evidence for the existence of other solar systems stems from a number of sources: current theories of stellar formation that hypothesize that stars and planets are created about the same time, measurements of the gravitational perturbations of nearby suns by unseen objects (that could turn out to be planets), and studies of double-star orbits (in which twin star systems are scanned for orbital irregularities that might indicate the presence of planetary objects). It's true that some of the stars in the heavens are too small or too unstable to support planets. However, these anomalies (dwarfs, exploding stars, and so on) are few and far between; most stars behave roughly like the earth's sun and have about the same longevity.

Evidence for planets in other stellar systems has, in fact, been mounting in recent years. In 1991, Matthew Bailes, Andrew G. Lyne, and Setnam Shemar, three British astronomers at the University of Manchester, reported in the science journal *Nature* that they had discovered a new planet orbiting a distant pulsar. A *pulsar* is an extremely dense stellar body, called a *neutron star*, rotating in such a way that it emits radio waves in pulses. These signals, normally spaced at regular intervals, were found in this case to be irregular, providing a strong indication of an orbiting object of planetary size. The 250-foot-diameter Lovell radio telescope at Jodrell Bank, England, recorded these erratic pulses about three times a second and determined the planet's distance to be 2×10^{17} (two hundred thousand trillion) miles away. Its size was found to be two or three times larger than that of the earth, while its mass was calculated to be about ten times that of our own planet. Furthermore, its orbital radius was estimated to be roughly the same as that of Venus around our sun. This monumental discovery, along with other new findings, indicates that it is almost certain that there are numerous distant planets in our galaxy.

Assuming then that each solar system contains about the same number of planets as ours (ten, give or take a few), that means that there are approximately 4×10^{11} planets in the Milky Way. In addition, many of these worlds may themselves be orbited by large satellites that are inhabitable in their own right. Almost certainly there are more than 100 billion planets or satellites buried within our galaxy just waiting to be explored.

Not all of these planets would be interesting to those searching for life-forms; the majority would be uninhabitable because of hostile conditions. In our own solar system, as we now know, only our own planet is fully suitable for life as we know it. Other worlds, such as Mars and Titan (a satellite of Saturn), have some of the prerequisites for life but lack such essentials as abundant water and sufficient oxygen. So, let's be conservative and estimate that only one of every fifty planets in our galaxy is inhabitable. This would still leave about 10^{10} planets—10 billion worlds that could support the simple necessities of organic existence.

In recent years, conclusive evidence has shown that genetic material and proteins could be formed on a previously lifeless planet in a primordial "soup" of carbon and nitrogen-rich nutrients. What would be needed for life, in addition, is plenty of energy in the form of solar radiation and electrical activity. Then, from the primitive strands of ribonucleic acid (RNA) and deoxyribonucleic acid (DNA), replication could lead to the development of more and more complex forms of living matter. Finally, after millions of years, large-scale life itself (fish, mammals, and so forth) could evolve.

These findings suggest that the formation of life would be reasonably common on planets with the right conditions. If only one of every three of these planets that can support life actually contains life, this would still leave a massive number of worlds out there teeming with organic activity: more than 3 billion ecospheres.

It seems reasonably likely that advanced life, once formed, would evolve in a manner similar to earth's species. Therefore, many of these ecospheres might also contain intelligent life: civilizations ready to be discovered, cultural systems waiting to be investigated. There may, in fact, be hundreds of thousands of planets in our galaxy already supporting advanced societies beyond our imagination. Curiously, none has contacted us; perhaps this is a function of the tremendous distances involved.

With the nearest inhabitable planets perhaps hundreds of trillions of miles away, we surely could not explore these distant worlds by present technologies. On the other hand, the existence of a Star Gate, such as the one described in *2001: A Space Odyssey*, would change this picture considerably. Clearly, the discovery or creation of a cosmic gateway would make it possible for us to reach the numerous planets beyond our current capabilities. Almost certainly we would find life out there, perhaps even intelligent life. The universe would lose its baffling anonymity and its unsettling silence.

So the Star Gate certainly is an intriguing concept, but isn't it a fictional one, a figment of the imagination of science fiction writers such as Arthur C. Clarke? Quite the contrary—the existence of cosmic gateways is a topic that is presently under serious and intense debate from California to Moscow. Over the last few years a number of serious proposals for interstellar passage have been published in major professional journals.

Theoretically it seems possible to design such a gateway, although the technical problems are formidable. Postulated objects called *black holes* and *wormholes* are said to possess the property that they connect two different parts of the universe. Thus, at least in theory, the era of the Star Gate has already arrived. And perhaps the unexplored parts of the cosmos will be surveyed sooner than we would have thought.

To understand these amazing and revolutionary developments in the field of astrophysics, as well as their implications, we must first explore the remarkable theory of space, time, and gravity proposed by Einstein in the early part of the twentieth century. Einstein's special and general theories of relativity continue to be considered the most accurate models of how planets move, how stars evolve, and how the universe expands. His ideas serve as a way of fully describing the overall dynamics of the cosmos in a way that is both elegant and profound. Furthermore, the Einsteinian concept of space is a malleable one, allowing for the existence of gateways: tunnels that link separate parts of the cosmos.

Thus, via Einstein's notions, we'll see how the right amount of matter, properly placed, might produce distortions in the paths of objects. In doing this, we'll be on our way to examining the possibility of the ultimate path distortion: the incredible spatial shortcuts provided by cosmic gateways, real-life Star Gates.

CHAPTER 2

COSMIC ROADMAPS

*L*et There Be Light

It's hard to imagine what it would be like to live in a world without light. Picture the darkest darkness that you have ever experienced: a crowded forest on a moonless night, the bottom of a deep chasm, a cellar with the lights off—any place that is murky enough to be disorienting. Chances are, you'll find that this vision contains some light; somehow you could make out faint images in the gloom. To eliminate all visible light you would need to place yourself in something like a sensory deprivation tank.

Even if you could block out all *visible* light, you still wouldn't live in a light-free world. For example, two blind men, though unable to perceive visual images, could still communicate with each other by means of radio. One of the men could speak into a transmitter, which would convert his sound waves into radio waves. The radio waves could be broadcast and picked up by the other man's transmitter, where these signals would be promptly converted back to sound waves.

These radio waves represent an invisible form of light, one that permeates the earth and heavens alike. The full range of luminal phenomena is enormous: light varies in type from radio signals of wavelengths an inch to a mile long, to infrared radiation (used in cooking), to the familiar visible spectrum (red through violet), to ultraviolet (which produces suntans), to X-rays (used in medicine), and, finally, to gamma rays of wavelengths a trillionth of an inch long.

A world without light would be a strange world indeed. Communication, cooking, solar energy, and X-ray tests would all be nonexistent. Yet even these common uses of luminous energy are but an infinitesimal part of the role that light plays in all of nature. It is no coincidence that the majority of the world's creation myths begin with some variation of the phrase "Let there be light." For without light, the earth as we know it would literally cease to exist.

According to the current quantum model of atomic structure, each atom contains a number of negatively charged electrons surrounding a nuclear core of positively charged protons and chargeless neutrons. When stimulated by external light waves of appropriate frequencies, some of these electrons may jump to new positions, either farther away from or closer to the core. Additionally, electrons may spontaneously hop to lower positions, as long as they obey what is called the *Pauli exclusion principle*: a sort of anticrowding ordinance for electrons (and similar particles) that prohibits more than two electrons from occupying the same distance from the nucleus. The Pauli principle describes a built-in mechanism that repels large numbers of electrons from the core and thus maintains structure for the atom. It serves the same function as a warning sign at a bus stop telling passengers to stand in line, two abreast; without it, they might all rush to the front of the queue at once and create chaos. Similarly, the exclusion rule, by demanding that electrons pair up, prevents them from all plunging at once into the innermost regions of the atom.

Each time an electron moves to a closer location, relative to its original distance from the nuclear core, it emits a photon (particle of light). Whenever an electron moves farther away, it absorbs a photon. Thus, for most atoms, light is radiated and absorbed fairly often—each time an electronic transfer takes place.

All chemical reactions are dependent on these electron and photon interactions; hence, without light, chemistry itself would be impossible. And, of course, it is these chemical processes that have built up the various molecules that constitute most of the materials with which we are familiar. Furthermore, light plays such a basic role in vital subatomic processes that even matter's constituent particles themselves, for the most part, could not have been created during the early stages of the universe without the existence of luminous energy. Hence the invocation "Let there be light" implies "Let there be earth."

Even before the discovery of the atom, it was well known that the oscillating motion of charged particles, such as electrons, caused light to be produced. The English physicist James Clerk Maxwell in 1862 predicted the existence of electromagnetic waves, generated by the fluctuations and accelerations of charges. He calculated the velocity of such waves to be exactly the speed of light, namely 186,000 miles per second. Then in 1887 Heinrich Hertz was able to produce significant quantities of electromagnetic waves by inducing electrons to oscillate over a million times per second. By causing these waves to reflect off metallic surfaces in the same manner as light, he conclusively showed that visible light is simply a form of electromagnetic radiation.

Interestingly, Maxwell's equations of electromagnetism show that, in a vacuum, the velocity of electromagnetic waves appears the same no matter what speed the experimenter is traveling. Thus, according to Maxwell's theory, observers in a rocket ship, in an airplane, or just resting on earth would all determine the measured speed of electromagnetic radiation to be the same value. This implies, of course, that the vacuum speed of light is constant, a result confirmed by the careful measurements, also in 1887, of Albert A. Michelson and Edward W. Morley.

The discovery of the constancy of the speed of light in empty space was extremely disturbing to the turn of the century physics community. Never before, in the history of science, was a speed found to be an observer-independent constant. On the contrary, the time-tested principle of *Galilean relativity* predicted that, depending on the velocity of the viewer, the motion of any object would appear different.

Basically, the Galilean theory of relativity is a formal statement of the experience of two people passing each other with different velocities; their relative speeds, as measured by one another, would appear faster or slower than their "actual" speeds, as measured by an observer at rest. Imagine, for example, what would happen if a truck driver, moving west at 50 miles per hour, were to pass a taxi driver, moving east at 50 miles per hour. Clearly, the cab driver would observe the trucker to be moving relative to him at 100 miles per hour—twice the truck driver's actual speed. However, if both drivers were traveling in the *same* direction, their speeds relative to each other would be 0 miles per hour: that is, they would seem to one another to be at rest.

However, this situation clearly couldn't apply to light. According to Maxwell's theory, as well as Michelson and Morley's experimental results, the speed of light is observer-independent. A trucker zooming by a beam of light at an incredible speed would still observe the light particles to be moving at 186,000 miles per second, even if he managed to travel at this speed himself. He could never catch up to the light, no matter how fast he drove; the light would always appear to be "fleeing" him at the same tremendous rate.

Because of paradoxical situations such as these, for physicists of the late nineteenth century it was abundantly clear that there was a contradiction between the principle of Galilean relativity and the notion that the speed of light in empty space is constant for all observers. Yet no one knew how to resolve this puzzling dilemma—no one, that is, until a young German-Jewish patent officer residing in Switzerland decided to take time off from his job to scribble down some notes about this problem.

*T*rains of Thought

Albert Einstein was born in Ulm, Germany, in 1879, eight years before Michelson and Morley's important experiment. Several years after his birth, the Einsteins moved to Munich and young Albert was sent to school. Growing up with an insatiable curiosity about the nature of the physical world, he was frustrated by the regimented atmosphere of his schooling and spent a great deal of time learning science and mathematics on his own. Finally, he dropped out of high school at the age of fifteen and headed first to Italy and then to Switzerland, where he enrolled at the famous Zurich Federal Institute of Technology (ETH).

At the age of twenty-one Einstein graduated from ETH and obtained work teaching at a technical high school. However, he was soon disappointed there by the lack of opportunities to teach at a more advanced level. These feelings soon became known and Einstein lost his job.

Unemployed, Einstein enlisted the help of a friend of the family, who found work for him at the Swiss Patent Office in Berne. Einstein's new position as patent examiner consisted of inspecting blueprints for inventions to determine whether or not these devices represented useful innovations. This task left him plenty of time to pursue his own hobby, performing detailed calculations involving key outstanding problems in theoretical physics. He would scribble his notes on a piece of paper, which he would hide if he heard anyone coming.

In 1905, Einstein's theoretical work finally came to fruition and he had four pivotal research papers published. One of these articles, concerning the so-called photoelectric effect (the principle that operates electric eyes), later won him the Nobel Prize. Two of the papers involved the nature of molecular properties. But it was the fourth research paper, concerning the special theory of relativity, that earned Einstein the universal acclaim that he enjoys to this day.

The special theory of relativity was Einstein's solution to the dichotomy between the principle of the constancy of the speed of light in a vacuum and the Galilean principle of relativity. Einstein knew that in order to solve this problem a fundamental change was necessary in the way space and time were considered. Clearly, the Newtonian view that space and time are absolute, which held that all observers would measure identical spatial distances and temporal durations between the same pairs of physical events, needed to be reconsidered and ultimately replaced by a new set of axioms.

Special relativity can best be understood by use of thought experiments developed by Einstein himself. Einstein was fond of railways and based many of his examples on situations involving moving trains. This gave a certain refreshing element of realism to conceptual dilemmas that might have seemed otherwise far more abstract and inaccessible.

The first thought experiment concerns a rethinking of the notion of simultaneity. Two events are said to be simultaneous if they are observed to occur at exactly the same time. One might think at first that this concept wouldn't depend at all on the motion of the viewer, and that simultaneity of occurrences would be agreed upon by all observing parties. However, Einstein's thought experiment shows that this assumption of absolute simultaneity must be considered false in order for the constancy of the speed of light to be upheld.

Imagine a very long train that is traveling at an enormous speed (close to that of light) past a railway station platform. Two observers, Fred and Mike, are situated such that Fred is sitting inside the train at the exact center and Mike is standing on the stationary platform. Suddenly the train is struck on both ends by bolts of lightning. Because Fred is in the middle of the train, the light from both bolts takes the same amount of time to reach him, and he concludes that the bolts have hit the train simultaneously. On the other hand, the light signals emitted by the lightning bolts reach Mike at different times (a few minutes apart, say). This delay is due to the increased distance that one of the bolts must travel, compared to the other, because of the fact that the train is moving. Hence, Mike, in complete disagreement with Fred, would regard the striking as being *nonsimultaneous*. Rather than being an absolute, observer-independent notion, simultaneity is dependent on the speed of the observer.

The relativity of simultaneity, combined with the constancy of light speed, implies that temporal durations must also be relative. For instance, in the train example, Mike and Fred, comparing notes, could reach one of two conclusions. The first possibility is that the light from the two bolts has traveled at different speeds in Mike's case but at the same speed in Fred's case. This conclusion must be discarded, however, because it implies that light could travel at different velocities. The other possibility is that Mike's clock runs slower than Fred's. Clearly this second conclusion, that the rate of time flow depends on the relative speed of the observer, must be the correct one.

This special relativistic effect, called *time dilation*, can readily be seen by setting up a special sort of "clock," consisting of a flashlight, a mirror, and a light detector, on board the train. Imagine that the mirror is placed on the ceiling of the train and the flashlight and detector are placed on the floor directly below it, facing up. When the flashlight is turned on, its light beam is bounced off the mirror and picked up by the detector. The detector, in turn, registers the time taken for the entire process. On the round-trip of the light beam, the detector ticks off one unit of time. Let's say, for the sake of argument, that the entire trip takes one second (in reality it would take only microseconds).

We consider now the case of two onlookers, Fred and Mike, one sitting in the moving train next to the detector, the other viewing the same device from the perspective of the fixed platform. Clearly, according to Fred, the moving observer, the time registered by one tick of the clock would be one second. He is sitting close to the detector, so naturally the speed of his watch would correspond to the pace of the device.

On the other hand, it's easy to show that Mike, the fixed observer, would note the time taken for one tick of the device to be longer than one second. Because of the train's horizontal motion, the light would appear to him to be traveling in a nonvertical, zigzag manner. Hence, from Mike's point of view, the beam would seem to be moving a greater distance between ticks than it would have if it had purely vertical motion. Since the speed of light must be exactly the same for Mike and Fred, the measured time would be slower for the former than for the latter: that is, for Mike there would be a longer duration between ticks. This increase in duration could be precisely measured: it would turn out to be a direct function of how close the train's speed is to that of light.

Time dilation and relativity of simultaneity are two of the most important special relativistic effects predicted by Einstein; both have been confirmed by laboratory experiment. Length contraction is another relativistic property that can be demonstrated by thought experiment and verified in the lab. This principle states that if an object were to travel close to the speed of light, it would appear, to a fixed observer, to shorten in the direction of motion. In other words, a yardstick, moving at near-light speed, would seem to be less than thirty-six inches in length.

Finally, in a fourth startling conclusion, relativity predicts a noticeable increase in the mass of a fast-moving body. It turns out that this mass increment results from the increased energy of motion of the object. This amount of energy can be shown to approach infinity as the object's velocity approaches that of light. Hence, the speed of light could never actually be reached by an object, since it would clearly be impossible to transfer an infinite quantity of energy to a moving body in order to accelerate it to light's velocity. As with Einstein's other results, this relativistic mass increase prediction has been confirmed in the laboratory.

Although Einstein derived all of the basic principles of the special theory of relativity by 1905, it wasn't until 1908 and the work of Hermann Minkowski that it coalesced into a unified model of physical reality. Minkowski, one of Einstein's former teachers, had a unique geometric perspective on the theory; he treated space and time as different manifestations of the same entity, which he called *space-time*. By augmenting the three spatial dimensions, length, width, and height, with one temporal dimension, he found that he could reproduce Einstein's results with simple equations that would place space and time on almost equal footing.

The key to understanding Minkowski's work rests on three mathematical concepts: the space-time diagram, the metric, and the so-called Lorentz transformation. A *space-time diagram* is a four-dimensional graphical representation of reality that pinpoints each physical event according to its spatial and temporal coordinates. For example, suppose on New Year's Eve, 1993, Marcia is standing in Times Square, New York. Her space-time coordinates, mapped on the diagram, would be something like seventy-two degrees longitude, forty-one degrees latitude, one hundred feet or so elevation above sea level, and a time of midnight, December 31, 1993.

Once all events are plotted on a space-time diagram, the *metric*, a mathematical function representing all possible spatial and temporal intervals, serves to provide information about the "distances" between events. These distances are not determined by the Pythagorean theorem used in simple plane geometry, which states that the square of the distance between points is the sum of the squares of the length difference, the width difference, and the height difference. In contrast, the square of the *metric distance* is determined by the square of the time difference *minus* the sum of the squares of the length difference, the width difference, and the height difference.

This minus sign plays an extremely important role in the theory. Without this sign, space-time would be identical to regular, three-dimensional space, albeit with an extra coordinate, and the square of the metric distance could only be a positive number. However, with the minus included, there are three possibilities for the sign of the square of the distance: positive, negative, or zero. The nature of this sign, called the *metric signature* (signature, in this context, refers to a positive, negative, or zero value of the distance squared), provides a vital piece of information about the relationship between the two events under consideration.

If the signature is positive, then the separation between the events is said to be *timelike*. This means that a causal relationship is possible between the events: one occurrence might influence the other at a later time. For example, a solar flare at 12:00 might affect the earth's weather ten minutes later.

If the signature is negative, on the other hand, then the separation is said to be acausal or *spacelike*: there could be no relationship between the two events. A beetle crawling on Betelguese at 12:00 couldn't possibly hope to influence an aardvark on Arcturus at 12:01.

Finally, if the signature is zero, there is a *lightlike* (also called *null*) separation between the occurrences. In that case, the events are divided in space and time just enough to be linked by a light signal traveling between the two of them. In other words, their time difference is exactly the time light would take to span their spatial distance. The two ends of a laser beam, for instance, are necessarily separated by a lightlike displacement, since the signal linking them travels at precisely the speed of light.

The last component of Minkowski's formulation of special relativity concerns the means of depicting the relative motion of reference frames (points of view)—as in the case of the thought experiment involving the moving train and the fixed platform. Minkowski found that the use of a special mathematical device, known as the *Lorentz transformation*, can provide an exact geometric depiction of time dilation, length contraction, and the other effects predicted by Einstein. This notation can be shown to be equivalent to, but far more elegant than, Einstein's way of delineating the special relativistic equations.

*M*olding the Space-Time Putty

Minkowski's successful reformulation of special relativity in terms of a four-dimensional model of the universe was of enormous help to the increasingly famous Einstein. First of all, Minkowski's simplification of the theory served to make Einstein's ideas far more accessible and attractive to researchers. Consequently, Einstein's reputation spread among the scientific community and he was granted, in 1911, a position as full professor at the German university in Prague. In 1912, he received another professorship at the Swiss Federal Institute of Technology, and finally, in 1914, he gained an enormously prestigious position at the Prussian Academy of Sciences in Berlin. Finally he could devote all of his time to his research, undisturbed by other duties.

The second contribution that Minkowski bestowed on his former student was an even more important one. By reworking Einstein's theory, he paved the way for an entirely new approach to gravitation. Einstein saw in Minkowski's work a way in which the laws of nature, particularly those related to gravity, could be woven into the very structure of space and time itself.

If one views a three-dimensional depiction of *Minkowski space-time*, as the diagrammatic model used in special relativity is known, one notices that the spatial and temporal axes are all perpendicular to each other. Space can be depicted, in this easily visualized construct, as being a two-dimensional flat surface—in other words, as a tabletop. Time, in this depiction, juts out from this surface at a right angle. All spatial distances can be sketched directly on the table top: that is, they are completely horizontal in this portrait. Purely temporal displacements, on the other hand, can be depicted as entirely vertical line segments. Finally, space-time distances, as recorded by the metric, consist of diagonal lines involving both temporal (vertical) and spatial (horizontal) displacements.

Note that in this three-dimensional picture of Minkowski space-time a *flat* surface is used to represent space. This is no coincidence—Minkowski space-time is a simple variation of what is called *Euclidean space*: the ordinary, boxlike coordinate system used in high school geometry. Figures sketched in Euclidean space are mapped out using x, y, and z axes—all at right angles to each other. Similarly, sketches in Minkowski space-time use the perpendicular x and y axes, as well as t, the time axis. In the full, four-dimensional space-time portrait used to represent special relativity, the z axis, at right angles to the other three, is used as well.

For this reason, Minkowski space-time is also called *flat* space-time. It can be pictured as a stretched out, completely flat rubber surface: a tightly stretched trampoline. The space-time points, in this analogy, can be best visualized as evenly spaced white dots covering the entire surface in a checkerboard pattern. All of the distances between consecutive points on the plane are equal, in both the lengthwise and the widthwise directions. Consequently, the space-time metric consists of an array of identical lengths, each characterizing neighboring intervals on the surface.

What would the motion of a tennis ball be like on this flat trampoline? Clearly if a ball were rolled across the surface it would travel in an unswerving path along the plane. It would continue to roll in a completely straight line unless otherwise diverted. Naturally, this linear motion is due to the purely horizontal, smooth nature of the surface, as expressed by the evenness of the white dots and the uniformity of its metric.

Suppose, however, that an artist decided to transform the trampoline into a new example of modern art by carefully using needle and thread to sew links between some of the white dots. These string connections would pinch parts of the canvas closer together so that some of the points on it would become nearer to each other and others would become farther apart than before. Hence the trampoline's metric, representing the distances between neighboring spots, would be completely altered. What then would happen to the surface of the trampoline?

Obviously, if the white dots were no longer evenly spaced, because of being tied together by string, what would happen is that the previously flat top of the trampoline would then become rounded. Parts of the surface would deviate downward; others would become bumps rising upward. Thus, the rubber face of the trampoline would be completely distorted by the lack of evenness between the points. By altering the metric (varying the distance relationships among the points) a curved surface would be produced.

Now imagine what would happen if the tennis ball were rolled out onto the now-bumpy exterior of the trampoline. Instead of traveling in a straight line, it would sway from side to side and follow a curved path along the rounded surface. Clearly, the curvature of the rubber material would significantly alter the path of the round projectile.

This three-dimensional effect echoes the four-dimensional consequence of changing the space-time metric of a flat Minkowski geometry. A new set of distance relationships among points in space-time similarly produces a rounded four-dimensional "surface." This curved space-time is called a *Riemann geometry*, named for the German mathematician G. F. B. Riemann. Just as in the case of the tennis ball on the trampoline, objects moving in such a curved geometry naturally exhibit nonlinear motion—even if initially projected in a straight line.

Einstein was well aware of the unusual properties of Riemannian geometries; he had spent a great deal of time examining their attributes. When Minkowski showed that the theory of special relativity could be expressed geometrically in terms of flat surfaces, Einstein began to ponder the use of curved surfaces in expressing other attributes of motion not included by his earlier model. In particular, Einstein wished to use Riemannian geometry to incorporate the effects of acceleration and gravitation into a unified theory of projectile behavior.

In 1916, Einstein unveiled the results of many years of careful, detailed calculations regarding the nature of gravity: his general theory of relativity. By exploiting the properties of Riemannian geometry, he managed to find a way of reproducing and augmenting Newton's theory of gravitation in a manner consis tent with relativistic notions. Later experiments were to show clearly that Einstein's results were even more accurate than Newton's renowned conclusions.

Basically Einstein's general theory of relativity involves a careful set of relationships between the geometry of a region of space-time, as expressed by its metric, and the mass contained within the same region. All of these factors can be shown to be related by a set of mathematical quantities with special transformational properties, called *tensors*, that are dependent on the mass and geometry of the area.

Suppose one examines a typical piece of space-time— think of it as a segment of the surface of a trampoline. One could certainly write down the metric for this region, depicting all of the distance relationships among the points, and hence the curvature as well. The space-time metric constitutes one of the important tensors under consideration. One could also produce a detailed delineation of the mass and energy in the same section. This mathematical description constitutes a second relevant tensor.

Einstein showed that one can write down a simple series of equations relating these two important mathematical objects—namely, the metric and the mass-energy tensors. These relationships, called the *Einstein equations*, constitute a connection between the geometry of a region and its mass distribution. Hence the arrangement of matter in a certain area precisely dictates the curvature of the sector. This curvature, in turn, determines the trajectories of all objects in the region. As Misner, Thorne, and Wheeler, authors of *Gravitation,* an important text on this subject, put this: "Matter tells space how to curve. . . . Space tells matter how to move."

General relativity supplants the Newtonian law of gravitation. This earlier theory, successfully utilized for hundreds of years, involves the idea that two massive objects attract each other with a long-range force that is proportional to the product of their masses and inversely proportional to the distance between them. Newtonian gravitation is remarkably accurate for all practical purposes; however, detailed measurements taken during solar eclipses show that Einsteinian gravitation is the more precisely valid theory.

What constitutes this measurable discrepancy between the older and newer gravitational theories? Essentially, the difference has to do with the behavior of light. Newtonian theory predicts that gravity should not have any effect on light because photons (light particles) are massless. Massless objects do not attract each other, nor influence each other's trajectories, according to this outdated viewpoint. On the contrary, light is influenced by gravitational effects. The solar eclipse experiments have proved that massive stars deviate light beams in a precisely measurable manner, exactly as Einstein's theory predicts.

One can readily see why, in Einstein's theory, light rays are distorted by the presence of matter. It turns out that photon paths are altered for exactly the same reason that the trajectories of other objects are deviated. To help us to visualize this situation, we return to the trampoline analogy.

Consider, once again, that the four-dimensional fabric of space-time is represented by a stretched elastic trampoline (without threads). Normally, in the absence of matter, general relativity mandates that space-time is flat (Minkowski). Similarly, if no massive objects rest on top of the trampoline, the rubber surface neither sags nor bulges: it remains completely level. A tennis ball rolled on the plane naturally travels in a completely straight path, in the same way that light particles, planets, stars, and other objects follow straight line motions through Minkowski space-time. There is no reason for bodies to deviate from the shortest routes possible, and in the case of flat regions the shortest paths ordinarily consist of perfectly straight lines.

Now imagine that a large stone is placed on the trampoline. Naturally, the rubber surface sags. The trampoline is no longer flat, but instead possesses a significant curvature. Consequently, a tennis ball pushed onto its exterior naturally moves in a curved path. Depending on the initial speed and position of the ball, it either follows a parabolic path (like a comet that never returns to its original position), orbits around the rock, or spirals in toward it.

In a similar manner, the presence of a massive object, such as a star, planet, or moon, in a particular region of the universe causes the nearby space-time to distort. Mathematically, this corresponds to a change in the space-time metric representing the region. Distances become stretched and durations become extended in the direction of the distorting matter.

Suppose, then, that a celestial body or other such object is propelled into this region of nonzero curvature. Depending on the projectile's initial velocity, one of three possible sorts of motion takes place: it either moves in a parabola about the distorting matter, engages in an elliptical orbit around the matter, or spirals inward and collides with the massive, curvature-generating object. The first case corresponds to the case of a nonreturning comet; the second, to that of an orbiting planet; the third, to that of a crashing meteorite. Hence, the entire range of motion displayed by celestial bodies can be exactly modeled by general relativity.

Finally, consider the case of light. Light always follows the shortest possible path between two points, which, in the situation of flat space-time, is perfectly straight. However, the presence of a massive object in a space-time region distorts the very nature of "straightness" in that area. The metric, the primary mathematical object containing information about distances, is altered significantly by the heavy matter. Hence, the straightest lines possible for light to follow there are still curves, and light beams must engage in curved motion while traveling through the distorted region. This is one of the most pivotal predictions of general relativity.

The Schwarzschild Solution

Within months of Einstein's publication of his new theory of gravitation, the first important exact solution to his equations of general relativity was reported. In 1916, the German astronomer Karl Schwarzschild produced results corresponding to the curvature of space-time near a spherical mass. His solution, called the *Schwarzschild metric*, was found, in the case of solar or planetary size masses and fairly large distances from the center, to yield exactly the same behavior as Newton's earlier prediction for gravity's magnitude—namely, an inverse-square dependence on distance from the center of the material.

The Schwarzschild solution is the four-dimensional equivalent of the results of dropping a spherical rock on a flat trampoline: a bulge in the cosmic fabric. Its metric represents a spherically symmetric disturbance (appearing the same in all radial directions, like the rays of the sun) that emanates outward from the mass. The magnitude of the distortion is directly dependent on how massive the central body is.

Schwarzschild's metric can be used to model the behavior of the solar system. By distorting the space-time in its vicinity, the sun causes the nine planets to move in elliptical orbits. Unlike in the Newtonian case, though, Einsteinian theory predicts that these ellipses must themselves rotate about the sun, albeit at a much slower rate. This effect, called the *advance of the perihelion*, has been experimentally verified in the case of Mercury, the only planet for which the theory predicts a large enough procession to be measured. Besides this deviation, however, there is little difference between the Schwarzschild and traditional models of the solar system.

Suppose, though, that one were to examine the region *inside* the sun. There, one would find quite a large discrepancy between the Einsteinian and Newtonian predictions. This deviation would begin to be more and more noticeable if one were to approach an imaginary distance from the center of the sun, called the *Schwarzschild radius*.

This radius, a function of the mass of the sun or other object under consideration, defines a spherical region for which general relativistic effects are extremely strong. Within its boundary the space-time curvature is extraordinarily large and path deviations are enormous. For the constituents of the solar system, these radii are relatively small—the sun's Schwarzschild radius is roughly one mile, the earth's, about an inch. For very large stars or galaxies, however, these radii are significantly greater.

The definition of the Schwarzschild radius concerns the property of a body known as its *escape velocity*. This is the minimum speed at which a nearby particle might break away from the body's gravitational influence. For example, the escape velocity at the surface of the earth is the least possible speed for a rocket ship to blast off from earth, namely, 6.8 miles per second. Naturally, for a body of greater mass and/ or for radial distances closer to its center, the escape velocity is larger—if the earth were four times smaller but just as massive, or just as big but four times as massive, its escape velocity would be twice as fast at 13.6 miles per second. Finally, for a particularly large mass and a significantly small radius, the escape velocity equals the speed of light. The value of the distance from the center at this point precisely defines the Schwarzschild radius.

The sun's Schwarzschild radius is so tiny compared to the total solar size that it's impossible to discern any of its effects. However, imagine shrinking the sun down to a tiny ball of less than one mile in diameter, while keeping its mass constant. This, of course, would render the sun incredibly dense. In that case, the Schwarzschild radius would lie *outside* the sun and the effects of traveling near it would be immediately noticeable. The space-time curvature within the confines of this sector would be extraordinarily great and would approach infinity near the center. Furthermore, the escape velocity for particles closer to the sun than the Schwarzschild radius would then be greater than the speed of light.

It is fascinating to consider what happens when a massive object is dense enough that its Schwarzschild radius is physically penetrable, lying exterior to the object itself. For an

object of solar dimensions, its density would be enormous
—over 10^{16} times that of water; in order to conceive of an
object with water's density the size would need to be over
10^8 times bigger. In any case, if the escape velocity near a
star's exterior were to exceed that of light, not even photons
themselves could escape from the star's surface. Light shone
on such a strange structure would be completely absorbed
by it; hence the object would appear at all times to be totally black. It is for this reason that we call celestial bodies
that have shrunk or have been compressed below their
Schwarzschild radius *black holes*.

A Schwarzschild black hole represents a sort of "bottom-
less pit" in space-time. The curvature in its interior is so
great that it prevents all internal objects from moving in
paths other than inward spirals toward its center. To return
to the trampoline analogy, it would be as if a giant rock were
to drop on the trampoline's surface and puncture the material, creating a large break through which other objects
could fall. Similarly, in general relativity, a black hole embodies a gap in the very fabric of the universe.

Not only does the inside of a black hole behave strangely,
the immediate exterior is anomalous as well. The "pitlike"
nature of the region directly outside a Schwarzschild radius
causes the time scales of objects in the vicinity to stretch
out. For clocks, this means that they would appear to slow
as they approached the boundary. For light waves, on the
other hand, the stretching out of temporal durations as they
approached would cause their frequencies to lower significantly—a phenomenon known as *redshifting*. The whole
spectrum of these rays would alter, oranges becoming reds,
greens becoming oranges, and so forth. Thus, a gold pocket
watch falling into a black hole would appear slower and
slower—and redder and redder—as it neared the invisible
border.

What would actually happen if an object were to cross
over the Schwarzschild radius of one of these models? Ever
since these enigmatic solutions of Einstein's equations were
found, scientists have been pondering this intriguing question. It was originally thought, because of some mathematical
difficulties associated with this problem, that the boundary

zone between the exterior and interior constituted something called a *singularity*: a point at which cause and effect breaks down and space-time reaches an impenetrable barrier, that of infinite curvature. In other words, at first it was believed by many that the Schwarzschild radius formed the very edge of space and time itself.

Later, physicists revised their forecasts about the nature of the boundary layer. In 1960 the American mathematician M. Kruskal and the Hungarian G. Szekeres independently introduced a new system of coordinates that removes the apparent singularity and successfully bridges the gap between the exterior and interior regions. In their innovative construction, they found that, instead of being a sort of barrier, the Schwarzschild radius serves as an invisible one-way boundary, called an *event horizon*, with very special mathematical properties. It turns out that light and other objects can cross the event horizon in one direction, but not in the other.

The results of this work are best embodied in a special graph, called a *Kruskal diagram*, which looks like a giant X. The two line segments forming the X represent here the event horizons dividing exterior and interior sectors. Clockwise around the X there are four different "domains," the right, bottom, left, and top, corresponding to four separate coordinate regions. The rightmost domain, labeled I, and the leftmost domain, labeled IV, represent the outside world— the exterior of the event horizons. The fact that these are two separate, disconnected regions is a matter of considerable controversy; various theories assert that section IV is either another part of our own universe, a different universe altogether, or perhaps just a mathematical fluke (the latter possibility would be most alarming, as it would indicate that general relativity does not provide a complete portrait of the cosmos). Later, when we return to the Schwarzschild model to examine the question of interstellar gateways, I'll discuss this situation further.

Regions III and II, the top and bottom parts of the X, respectively, correspond to the interior of the Schwarzschild model. The third domain, called the *interior future*, is a region to which signals can freely travel (from domains I and IV) but from which they can never be emitted. Hence, if Mandy is located in region III and Sandy stands in region I, Sandy might contact Mandy, but Mandy could never respond to her. Thus, the event horizon between regions I and III acts as a one-way hindrance to communication.

The second domain, called the *interior past*, is the exact opposite of the interior future: signals can leave this region but can never enter. Hence, one-way communication is possible from region II out to regions I and IV, but not vice versa.

The Kruskal diagram is a bit like a blood donation chart. The exchange rules are quite similar: Blood type O (read: domain II) can donate to types A and B (read: domains I and IV) but cannot receive from these types. In turn, types A and B can donate blood to type AB (read: domain III) but cannot receive from AB or from each other. The "event horizons" in these cases prevent "two-way donations" from being possible.

Consider what happens when an object passes from region I to region III. Once inside the event horizon, no return to the outside world is possible; nor is any outward communication. The object is drawn inextricably toward the deepest interior, represented on the diagram by the very top of the **X**. Similarly, passage to any of the other regions is also impossible. Thus, the event horizon located at the Schwarzschild radius truly represents a "point of no return."

What does the top of the **X** represent? This segment, the physical center of the Schwarzschild solution, consists of a real, unavoidable singularity—an abrupt break in the continuity of time and space. All bodies moving through region III eventually reach this singularity, where they are crushed into nonexistence—they literally fall off the edge of spacetime. Nothing can be given off by this singularity—and everything in its vicinity is destroyed. Hence this interior future singularity can be said to be a perfect "consumer."

A second singularity of the Kruskal coordinates is located at the very bottom of the **X**, in region II. This is a point from which all paths in the region emanate. Although matter can be produced from this singularity, no matter can be gobbled up there. This singularity of the past, a perfect "producer," is sometimes referred to as a *white hole*, the implications of which will be discussed in Chapter 4.

The Kruskal diagram can be used to model other solutions of the Einstein equations. Although the Schwarzschild model is electrically neutral and nonrotating, charged and/or rotating astrophysical models have also been found. In 1963, Roy Kerr, a New Zealand physicist, derived the solution of Einstein's equations for a neutral, rotating, massive object: the *Kerr metric*. This was extended in 1965 to include the charged case, usually called the *Kerr-Newman model*.

In each of these cases, Kruskal diagrams can be produced, showing all of the event horizons, singularities, and interior regions. It turns out that the diagrams representing massive, rotating objects—whether charged or neutral—depict a far more complex setup of regions and boundaries than those describing the Schwarzschild example. The diagram for the Kerr-Newman solution, for instance, looks like an infinite linked series of X's attached to each other in a paper-doll fashion. Event horizons enable signals to pass unidirectionally between some neighboring regions; singularity barriers prevent other sorts of communication.

It is interesting to ponder what this endless chain of sectors represents. Although some skeptical scientists discard any solution that includes the possibility of "other worlds," viewing it as smacking too much of science fiction and UFOs, many theoretical physicists now believe that this infinite collage of regions maps out a detailed set of connections between distant sections of the universe. General relativity has proved correct in all cases so far; why doubt its implications when it prophesies a multiply connected cosmos? Certainly, the picture of astrophysical models depicted by Kruskal diagrams lends a considerable amount of hope for the possibility of cosmic shortcuts.

Interstellar Shortcuts

Today, all but a few renegade physicists accept Einstein's theory of general relativity as the most complete model of large-scale gravitational effects in the cosmos. Yet not every one of these scientists agrees with one of the most radical conclusions of this theory: that there exist tunnel-like connections between otherwise separate parts of the universe. Clearly, though, several of the most important solutions of the Einstein equations that have been found embody two or more separate regions of space-time. And, without a doubt, this potpourri of worlds must have at least some physical significance.

The flexible nature of Einsteinian space-time, in contrast to the rigidness of Newtonian space and time, seems to permit the existence of geometries distorted greatly enough to allow for transuniversal links. According to the time-tested conclusions of theoretical general relativity, rips, tears, and connections in the cosmic fabric do actually exist. The presence of these geometric formations fuels considerable speculation that traversable interstellar shortcuts, connecting widely separated parts of the universe, may, in fact, be possible.

In the past few decades, an exhaustive search for these shortcuts, among the set of all solutions to the Einstein equations, has begun in earnest. A host of theoreticians have embarked on a quest for realizable methods of transgalactic transport, via cosmic links found in general relativity. So far there have been a number of dead ends, and several extremely promising leads.

Could star gates be created out of black hole links? The Schwarzschild model of nonrotating black holes seems to contain a link between two separate parts of the cosmos— or maybe even two different universes altogether. Might an astronaut successfully travel through this cosmic connection? On the other hand, it seems that the peculiarly complex nature of the Kerr and Kerr-Newman solutions of Einstein's equations of general relativity implies that rotating, extremely massive, gravitationally collapsed objects would generate an even greater array of cosmic connections; would any of these Kerr tunnels be navigable? The next chapter will explore the hitherto unsuccessful search for interstellar passage via black hole connections.

CHAPTER 3

BLACK HOLES: INTO THE VOID

*D*eath of a Star

The idea of a black hole is now so familiar that it's hard to believe that the term was coined only about twenty-five years ago by John Wheeler. Wheeler, a former Princeton University professor, is considered the granddaddy of modern astrophysics; the list of his former students and co-workers could fill a *Who's Who* of stellar theory. Before he lent his support to the notion that black holes really exist, these objects were viewed as mere glitches in the otherwise sound mathematical structure of general relativity. Wheeler's enormous influence brought about the recognition of black holes as an integral part of the theory of stellar demise.

Yet the origins of the concept of an opaque star reach back much further. Remarkably, it was over *two hundred years ago*, in 1784, that the first paper speculating on the existence of black holes was published in the British journal *Philosophical Transactions of the Royal Society*. Written by John Michell, this seminal work details a process by which a very massive star might lose its ability to emit light, because of the extraordinarily high escape velocity for such emission to take place. Light given off by such an entity would not have the speed needed to flee its orbit and would simply fall back to its surface. Hence, Michell argued, such an object would be completely black.

Ten years later, another work appeared in France, written by Pierre Laplace, independently speculating on a similar construct. In the original editions of his highly influential *The System of the World*, Laplace includes a reference to the notion of "dark bodies": objects large enough to capture all emitted light. Curiously, this remark was deleted from later editions, perhaps as a result of a new-found cautiousness of Laplace or his editors.

By the nineteenth century, allusions to such invisible bodies essentially died out, mainly because of an increasing awareness that light was wavelike and, so it was thought, incorporeal. Hence, according to Newtonian physics, it couldn't be affected by gravity. This lack of interest in the idea of black holes continued until the 1910s, when the formulation of general relativity, and the derivative notion of gravitational light bending, made it once more possible to talk realistically about the recapturing of luminous energy by massive stars.

It would seem that the discovery of the Schwarzschild solution to the Einstein equations would have spurred a massive search for Schwarzschild black holes as possible relics of the final stages of stellar development. Yet this was not the case: most scientists considered the light-gobbling limit of Schwarzschild's result to be a mere aberration. It was believed that the complex process of star death would lead to a far more complicated end state than the Schwarzschild metric, or any other simple solution of the Einstein equations, could model.

It took more than forty more years—a period that represents a pivotal stage in the emergence of modern stellar theory—for astrophysicists to resume pondering the physical meaning of black holes. This process was initiated, for the most part, by the work of two brilliant men who were each to win the Nobel Prize later in life: the Indian physicist Subrahmanyan Chandrasekar and the Russian scientist Lev Landau.

In 1928, the young Chandrasekar, while on his way to Cambridge, England, to study with a leading astronomer, Arthur Eddington, began to work out the details of a problem concerning stellar collapse. His task was to determine under what circumstances the end states of stars are stable. It was reasonably well known at the time that stars burn hydrogen fuel in a thermonuclear process which generates a great deal of excess energy. The enormous outward pressure of this additional heat balances the substantial inward gravitational force of the stars' constituent atoms and helps to prevent these stellar bodies from caving in on themselves. Chandrasekar wondered what would happen when the hydrogen and other nuclear fuels were exhausted. Surely a collapse would take place. The question was, Would the contraction reach a limit or would it continue until the star was completely reduced to a point?

Chandrasekar calculated the influence of the primary factor that could prevent a star from complete collapse, the repulsion caused by the Pauli exclusion principle. He found that for relatively small stars, the exclusion principle would keep their electrons sufficiently separated to block the stars' gravitational collapse. However, for stars more than about one and one-half times the mass of the sun, a mass now known as the *Chandrasekar limit*, this force of resistance would not be enough to support them against further shrinkage. Hence, according to Chandrasekar, there were two possible stellar end states, depending on their initial mass: a stable tiny star, known as a *white dwarf*, or a physical state far more compact than ordinary matter, that is, what came to be called a black hole.

About the same time that Chandrasekar was completing his studies, Lev Landau was independently reaching similar conclusions about the nature of stellar demise. However, Landau postulated *three* conceivable end states for stars, depending on their mass: white dwarfs, the so-called neutron stars, and black holes. The densities and other properties of these final forms are determined by how much resistance the internal forces, such as the Pauli exclusion principle repulsion, display against gravity—with the white dwarfs maintaining the strongest resistance and the black holes, the weakest.

Let's now consider these three different stellar decay scenarios sketched by Landau. In the first case, involving stars of about one solar mass or less, gravitational collapse would be halted by electron repulsion, resulting in stable *white dwarf* end states. Thus, white dwarfs, though highly compact, would nevertheless consist of ordinary (atomic) matter.

It was well known at the time of Landau that many large stars, such as Sirius, have tiny, radiant white dwarf companions. In 1841, the astronomer Friedrich Bessel determined by careful measurements of peculiar motional disturbances that the star Sirius had an unseen companion. This entity, Sirius B, was the first white dwarf star to be discovered. Later, in 1896, a similar small object was found circling the star Procyon, implying that white dwarfs are relatively common. Finally, in 1915, Walter Adams proved that Sirius B and similar stars were extremely hot, as well as being very heavy and extraordinarily compact—exactly the white hole properties predicted by theory.

Indeed, white dwarfs are exceedingly small, dense, and fiery. These stellar remnants have radii of just a few thousand miles and densities of hundreds of tons per cubic inch. Because of their small sizes—their diameters are comparable to those of average-sized planets—they are invisible to the naked eye. Moreover, their temperatures are more than two thousand degrees hotter than that of the sun. True white dwarfs represent extremes among astronomically detectable stars.

The second possibility for stellar death, in Landau's scheme, involves the collapse of stars that are about one and one half times more massive than the sun. The resulting object formed, called a *neutron star*, is even tinier and more dense than a white dwarf. Neutron stars are characterized by a composition of entirely nuclear matter (i.e., pure neutrons).

How are neutron stars produced from ordinary stellar material? First, in the case of these massive objects, gravitational forces become strong enough to overcome electron repulsion. Consequently, all of the atoms that constitute the star become completely unstable and all of their orbital electrons decay toward their nuclei. Eventually, all of the electrons are completely absorbed by the nuclei and no more ordinary atoms are left. Finally, this process stabilizes—because of the Pauli repulsion of the neutrons making up the material

—and the star remains frozen in this undiluted nuclear state. The resulting object can be said to be composed of an extraordinary sort of material called neutronium.

Neutronium, as postulated, would be incredibly dense, with densities 10^{14} times greater than that of ordinary solids. A chunk the size of a thimble would weigh about 100 million tons. The sun, if compressed into such a state, would be only a few miles across, the earth but a few inches. Because of these tremendous densities, objects resting on the surface of such compact bodies would need to achieve velocities close to half of the speed of light.

In 1967, the first celestial body considered to fall within this category was discovered quite accidentally. The incident occurred in Cambridge, England, when the British astronomer Antony Hewish asked his graduate student Jocelyn Bell to examine the data taken from a new radio telescope that he had built. Much to her surprise, Bell found, buried among the data, a curious sort of astonishingly regular radio pulse. At first, she thought this periodic "blip" was the result of man-made interference. Then, she guessed that it was a signal sent by a distant extraterrestrial civilization.

After ruling out these possibilities, Bell and Hewish came to a rather revolutionary conclusion: they had discovered a rapidly rotating neutron star—a *pulsar*, they called it, because of its pulsating signal. This neutron star was spinning so rapidly that the frequencies of its emitted radio waves, rather than maintaining constant bands, oscillated within a fairly broad range of values.

In 1974, Hewish was to win the Nobel Prize for this first indirect "sighting" of Landau's predicted object (for some reason, Bell was excluded). So far, no *direct* observations of neutron stars have been possible because of their extremely small sizes. However, following Bell and Hewish's lead, many scientists have made other detailed indirect measurements of similarly pulsating bodies, all believed to be examples of neutronium-filled collapsed stars.

Neither white dwarfs nor neutron stars would be of much interest to those seeking interstellar shortcuts. In terms of their gravitational effects, they represent dents, rather than gashes, in space-time and hence do not allow for connecting links between distant parts of the universe. All tunnels require entranceways, and these stellar relics simply do not supply them. Nevertheless, the existence of these peculiar celestial bodies portends well for the astronomical discovery of other structures that do contain traversable cosmic gateways.

In the cases of both white dwarfs and neutron stars, a physical constraint halts the complete gravitational contraction of these objects. But what happens if the initial mass of the star is greater than the Chandrasekar limit and the force of gravity overcomes all resistance forces? Landau didn't have much to say about this third case; he considered the notion of complete collapse too peculiar to be realized. It wasn't until 1939—and the work of the American scientist J. Robert Oppenheimer—that this question was resolved.

Oppenheimer is best known for his secret work with atomic bomb development during the 1940s and for his subsequent harassment by the House of Representatives Un-American Activities Committee (HUAC) during the 1950s. Yet it was much earlier, and in the field of astrophysics, that he would make his most important contribution to science by solving the gravitational collapse problem.

In an innovative article about stellar death, Oppenheimer proved that if a star possesses a mass more than about three times that of the sun, its fate is far more turbulent than that of white dwarfs and neutron stars. His scenario begins at the moment that the last drop of the heavy star's life-giving nuclear fuel is exhausted. Deprived of its source of power, the star begins to shrink in on itself rapidly in a vain attempt to convert the gravitational energy gained by contraction into the force needed for sustenance. Though futile, this shrinkage continues as the star becomes more and more compact.

In the event that the gravitational collapse of the star is especially sudden, a series of catastrophic explosions then occurs. Shock waves, sent out by the colossal contraction, rumble through the peripheral material of the stellar relic and then propel massive amounts of gas into the heavens in a spectacular display of fireworks. The resulting "light show" would be seen for millennia as the hazy, colorful outcome of what is called a *supernova*. The Crab Nebula, first spotted by the Chinese in 1054, is, no doubt, the consequence of such a cosmic catastrophe.

In any case, whatever is left of the once-enormous star would then proceed on its ever-shrinking irreversible course of doom. If the remnant were less massive, its internal quantum dynamics, embodied in the Pauli exclusion principle, could rescue it from complete collapse and a neutron star could be formed. However, because of the overwhelming weight of the stellar relic in this case, its substantial inward pressure would bear down heavily on its supporting framework and overpower any remaining resistance to total decay. Even the seemingly impregnable internal structures of the elementary particles forming the remnant would be smothered by the sheer magnitude of the blanketing gravitational forces.

Finally, what would remain of the once-radiant star would be a dark, formless pulp—an ultradense cauldron of matter and energy. The gravitational force produced by its formidable presence would be strong enough to prevent even light from escaping its vicinity. As direct observation of this relic is impossible, its ghostly existence could be felt only by its distorting actions on the paths of nearby objects. In short, it would have become a massive, invisible maelstrom: a black hole in space.

The Black Hole Appears

Oppenheimer's account of the gravitational collapse of extremely massive stars and of the dark vortex that survives this decay seems remarkably similar to the light-capturing scenario depicted by the Schwarzschild solution of general relativity. Indeed, in 1967 the Canadian scientist Werner Israel conclusively showed that all nonrotating black holes resulting from stellar demise could be precisely described by Schwarzschild's model.

Interestingly, Schwarzschild's black hole template is extremely regular in its physical properties: spherical, symmetric, and uniform. Furthermore, any two nonrotating black holes are identical in their makeup. One might think that such a simple solution of Einstein's equations would be hardly adequate to model the end product of complex stellar collapse.

However, detailed calculations by John Wheeler and the Oxford mathematician Roger Penrose have demonstrated how the perfectly simple might have evolved from the exceedingly complex. Penrose, an expert in the analysis of abstract geometric and topological constructs, is a relative newcomer to the study of astrophysics. Nevertheless, he has made an extraordinary impact on the field, even inventing his own simplifying notation. Over the years, working in both Britain and America, he has pursued a wide range of endeavors—from proposing a reason for time's unidirectional arrow to devising his own way to tile a plane, from postulating intricate mathematical puzzles to debating the significance of artificial intelligence. His collaborations with astronomers and physicists such as Wheeler, E. T. Newman, and Stephen Hawking have proved highly fruitful for all parties. In 1988, along with Hawking, he was awarded the prestigious Wolf Prize for their joint work in cosmology.

Penrose, with Wheeler, argued that the asymmetric, nonspherical components of a dying nonrotating star would be cast off during the star's death throes. All of the irregular elements would then be radiated away by the decaying object. The resulting remnant would necessarily be highly symmetric, perfectly spherical, and thereby entirely suitable to be modeled by the Schwarzschild solution.

So far we have considered only the *nonspinning* case. What then would happen in the more general situation of the decay of a *rotating* star? It turns out that this case permits a solution that is only slightly more complex than Israel's interpretation of Schwarzschild's work.

Basically, all stellar objects of more than three solar masses must eventually decay to forms identical to those mapped out by Roy Kerr in 1963 (the Kerr and Kerr-Newman solutions discussed in the previous chapter). These are the only possible sorts of black holes found in nature; Schwarzschild static black holes form a special subclass of these solutions. The Cambridge astrophysicist Brandon Carter sketched the rudiments of this important result in

1970; the proof was completed by the University of London theorist David Robinson in 1973 (with intermediate contributions by Hawking). Their work paved the way to a complete and highly successful theory of stellar demise and black holes that is still used today.

Carter and Robinson's work is often referred to by the moniker "A black hole has no hair." This strange expression alludes to the fact that all of the properties of a black hole are based on just three quantities: its mass, its charge, and its angular momentum (spin). All other pieces of information about the earlier form of the stellar remnant are lost during the star's collapse. Consequently, black holes display a curious uniformity—much like identically clothed U.S. Marines after their famous first-day closely cropped haircuts. As in the case of the Marines, the "no hair" condition of black holes mandates their almost complete similarity of appearance.

Black holes betray no trace of their origins. If captured, they could provide only three pieces of information about themselves: their mass, charge, and angular momentum number. No amount of careful prying could force them to reveal other secrets about their background. Hence the "no hair" theorem indicates that their pasts remain shrouded in mystery.

What sorts of backgrounds might black holes have to conceal? It turns out that these dark remnants may have been formed in a variety of ways, not just from the demise of massive stars. They may also have been created by the rapid collision and coalescence of large quantities of interstellar gas and material, as well as by the collapse of massive stellar systems or galaxies. In addition, the early universe may have contained billions of *primordial black holes*: tiny, ultracompact objects, as small as 10^{-5} gram, evolved from initial density perturbations in the cosmic fabric. Some of these entities might still survive; others may have formed the seeds of nascent galaxies.

Once born, black holes are ravenous creatures. They consume all matter and energy that happens to cross their paths and consequently grow in the process. Each time planets, asteroids, or other chunks of mass cross their one-way membranes (event horizons), these darkened devourers simply add bulk to their already bloated frames. Moreover, once they have finished their feast, they manage to destroy all evidence of their consumption via the information-annihilating "no hair" theorem. By guaranteeing that all that can be known about black holes are their masses, charges, and angular momenta, the "no hair" principle virtually ensures complete obliteration of all knowledge about black hole components.

Only woe comes to stars with black holes as orbiting companions. The fossilized associates use their privileged positions as binaries to launch into enormous feeding sprees, gobbling large quantities of the matter and energy produced by their still-vibrant neighbors. Eventually, they drain the remaining lifeblood, leaving them to spend the rest of eternity as shattered hulks—or, even worse, as more black hole fodder.

Black hole appetites are not restricted to interstellar dust and material drawn from other celestial inhabitants. No, these insatiable creatures sometimes even feed on their own kind. Larger black holes may cannibalize nearby smaller ones, while two or more such objects in close proximity may merge into a larger, more massive entity. Black hole growth by this method is virtually limitless.

The mechanisms and rules for black hole merger were largely discovered in the early 1970s by one man: the eminent British astronomer Stephen W. Hawking. Over the last few decades Hawking has made an inestimable contribution to the fields of astrophysics, particle physics, and cosmology. His patient work, conducted in spite of severe physical limits such as Lou Gehrig's syndrome (atypical amyotrophic lateral sclerosis), has revolutionized the study of modern gravitational theory in a manner second only to Einstein's. Although his body is confined to a wheelchair and his voice is restricted to a mechanical synthesizer, his mind continues to wander freely about the cosmos, solving many of the long-pondered mysteries of the universe.

During his graduate studies at Cambridge, Hawking was taken under the wing of the distinguished professor Dennis Sciama. Sometime during his first year there he contracted his degenerative disorder, but nonetheless he proceeded with great vigor on his research program in theoretical astrophysics. He became extremely interested in the problem of gravitational collapse and began, in the mid-1960s, to develop, along with Roger Penrose, key theorems proving that all black holes must contain unavoidable central singularities—points of convergence from which no light rays might escape.

About this time Hawking was appointed Lucasian Professor of Mathematics at Cambridge, the prestigious chair originally held by Isaac Newton. In his newfound capacity as full professor, he continued to work closely with prominent scientists of the United Kingdom, the United States, and Europe, including Carter, Sciama, Penrose, and Zeldovich (of Moscow), in developing a comprehensive theory of gravitational phenomena dealing primarily with issues revolving around black holes and the early universe.

Hawking's contribution to the problem of collisions between several black holes concerns the nature of the massive by-product that would be formed. First of all, he showed that if two black holes were to collide and coalesce, there would be no way of blasting the resulting object back into a pair of black holes. Hence all black hole collisions would be completely *irreversible*. Second, he proved that if two black holes unite, the surface area of the final black hole must exceed the sum of the surface areas of the initial black holes.

Simply put, Hawking's rules for black hole dynamics guarantee that they can only combine, not divide and grow, not shrink. Like the "Blob," the amorphous, voracious villain of the low-budget horror movie of the same name, black holes naturally increase in surface area as they consume and never lose the extra capacity that they've gained by this consumption. Thus, the total black hole portion of the universe is ever increasing.

Eventually, then, if the universe continues to evolve in the present manner and Hawking's laws hold true, black holes will form the bulk of the cosmos (if they do not already). These dark behemoths will then draw the last remaining stars and planets into their frozen interiors. In this bleak scenario, the universe will finally die a "heat death": a cold, gloomy demise. Fortunately, however, this process will likely take billions and billions of years to complete.

Cygnus X-1

Establishing the actual physical existence of black holes might seem, at first, to be an impossible task. Since black holes, by definition, do not emit any light, one might think that they would be as difficult to detect as "black cats in a coal cellar" (as Hawking puts it). Furthermore, because black holes of stellar origin would be so minuscule—roughly ten miles or less in diameter—and so tremendously distant—hundreds of trillions of miles away—at best they would appear as tiny dark specks against the mammoth sky. However, because these small proportions would mandate a visible angular size of about 10^{-8} second of arc, it's not at all clear how black holes could ever be directly observed.

An indirect method of detecting black holes that appears plausible at first glance involves their well known light deflection properties. Suppose a black hole were to cross the path of a distant star. Naturally, the black hole's gravitational pull would cause the light of the star to deviate from its ordinary course in a measurable way. The problem is that the chances of such a perfect alignment of star, black hole, and earthly observer are extraordinarily low, far too low for realistic expectations of success.

By far the most promising method of black hole detection involves the noticeable effects on visible stars of invisible binary companions. This could be achieved by carefully searching for stars seemingly orbiting around "nothing at all." Presumably, the masses of their unseen companions could be indirectly measured by careful observation of the parameters of orbital motion. The resulting values could then be compared to the masses expected for black holes. If the masses of the invisible entities were large enough, one might rightfully conclude that they represented unseen black holes.

This suggested approach is actually quite old: John Michell proposed it in his early (1784) paper as a way of detecting (what came to be called) black holes: "If any other luminous bodies should happen to revolve about them [black holes] we might still perhaps from the motions of these revolving bodies infer the existence of the central ones with some degree of probability, as this might afford a clue to some of the apparent irregularities of the revolving bodies. . . ."

Apparently, Michell predicted a successful modern experimental technique more than *two hundred years* in advance.

This procedure alone, however, would be far from complete. Although the method of binary star system observation would provide us with considerable evidence for black hole candidates, it would be insufficient to supply the vital proof needed to confirm that these dark objects behave in the manner predicted by gravitational theory. Other explanations for the invisibility of these bodies might be found. To supplement these orbital measurements, careful readings of the radiation emitted by the black hole candidates would be required. These emission readings would help to ensure the authenticity of these unseen objects.

The radiation given off by black holes in binary systems is mainly the result of energy created when hot stellar gases emanate from the visible star and plunge into the hidden companion. The total amount of radiation produced, a quantity astronomers call the *luminosity*, is a direct function of the amount of gas that falls toward the black hole and of the proximity of the two orbiting companions.

This luminosity would be very low for binary systems in which the visible star and black hole are spaced apart at typical distances, since the amount of gas passing between the objects would create very little energy. This would also be true for examples in which the visible stars were of average size, say that of the sun. Hence, this method would not work very well under most circumstances, because the light produced would be barely detectable.

When the intercompanion distances are very short and the shining star is very large, however, an enormous amount of gaseous material is transferred between the star and the black hole. This material swirls around the black hole, creating a nebulous formation called an *accretion disc*. As the gas from the disc spirals inward toward the black hole, large quantities of radiation are produced. Therefore, the luminosity of such a system is quite high and the prospects for detection of the emitted rays are very good. The radiation produced is mainly in the form of high frequency X-rays, emitted in irregular pulses. The irregularity stems from the erratic nature of the emissions as the stellar gases spiral toward the surface. These X-rays are relatively difficult to measure on earth and are far easier to detect in space.

It is for this reason that the *Uhuru* (meaning "freedom" in Swahili) X-ray satellite was launched in 1970. X-ray satellites contain special telescopes specifically designed to observe X-ray sources, such as matter falling into black holes. During its tenure, *Uhuru* detected 339 new X-ray sources, the majority found in 1972. These sources, once found, were then analyzed by instruments located on other satellites.

Several promising black hole candidates were discovered by these methods in the 1970s and 1980s. One such likely black hole was found in the vicinity of the constellation Cygnus. The X-ray source detected, called Cygnus X-1, was determined to consist of a gigantic visible star of twenty solar masses orbiting and being orbited by a massive invisible object of ten solar masses. The unseen body is far too heavy to be either a white dwarf or a neutron star; its mass is much larger than the Chandrasekar limit. Therefore, it is extremely probable that this object is a black hole formed from stellar decay. Additional evidence, including careful measurements of the X-ray luminosity and frequency range of the system, has strongly supported this point of view—that Cygnus X-1 contains a black hole of about twenty miles in diameter. Other likely black collapsed stars have been detected, by similar techniques, in the crowded center of our own galaxy, where massive stars are extremely common, and even as far away as the Magellanic Clouds (two neighboring galaxies).

X-ray satellites continue to be developed and launched with the hope of providing even more conclusive evidence of black holes. The Einstein Observatory (HEAO-2), put into orbit in 1978, was equipped with an X-ray telescope. In 1998,

NASA plans to launch a sophisticated new advanced X-ray astrophysics facility. It is hoped that this new installation, with its superior detection equipment, will help us piece together our now fragmentary understanding of black holes.

The Black Hole Disappears

After the discovery of Cygnus X-1 and other likely black holes, it became clear to most theorists that they had best start taking these objects very seriously. Spurred on by black hole "cheerleaders" such as John Wheeler, physicists largely abandoned their earlier apprehension about what they felt were mere theoretical aberrations in general relativity and began to grant much more credibility to the notion of compressed stars. By the mid-1970s, scientists started to concentrate more on incorporating black holes into the structure of mainstream physics and less on ascertaining whether they actually existed.

At that time, one of the most serious theoretical problems with black holes concerned their relationship to the field of thermodynamics. The century-old laws of thermodynamics, describing the relationships of concepts such as heat, temperature, and energy, were considered by theorists to be virtually irrefutable. Yet black holes seemed explicitly to defy one of these fundamental laws.

The problem rested not with the first law of thermodynamics, that of energy conservation, but with the second law, that of entropy increase. In the first case it was easy to show that black holes do not violate the principle that the overall energy of the universe, with all processes considered, must remain constant. Although black holes do absorb energy, in the form of light radiation and energized matter, they do not destroy it but merely convert it into other forms. Ultimately, the power sent into black holes is incorporated into their structures, in the form of increased mass.

However, black holes did seem to defy the second law of thermodynamics, which can be stated in a number of equivalent ways. One way of putting it concerns the somewhat abstract, but well-defined, physical quantity known as entropy. Entropy is the measure of a system's disorder; the entropy of a pile of sand is far greater than that of a sand castle, for instance. Similarly, the entropy of a tank of lukewarm water (a disordered state) is much greater than that of a divided tank, filled on the left side with hot water and on the right with cold water (a more ordered state, because of the barrier). According to the second law of thermodynamics, the total entropy of a closed system never decreases. That is why one wouldn't expect to see a dispersed pile of sand, without outside aid, evolve into an intricate sand castle. Nor would one anticipate seeing a tank full of lukewarm water spontaneously divide into hot and cold sections. Each of these incidents would involve a clear defiance of the law of entropy increase (more precisely, *nondecrease*).

It is easy to see how the second law of thermodynamics, as stated in this manner, produces a natural definition of the arrow of time. Time's arrow points in the direction of increasing entropy—of ever-growing disorder for all closed systems. That is why one could never encounter pure entropy-reducing events such as self-assembling sand castles; personal clocks would have to run backward for that to occur. Clearly, for forward-time creatures like us, the overall entropy of the world could never diminish.

That is why many scientists during the early 1970s were most perplexed by the energy- and matter-gobbling properties of black holes. By swallowing high-entropy chunks of the universe, black holes could seemingly be used as entropy-reduction devices. Simply by throwing large quantities of disordered objects, such as gaseous nebulae or interstellar dust, into the mouths of black holes, one could lower the overall entropy of the cosmos—perhaps even reverse the arrow of time—it was thought. This would represent a gross violation of the second law of thermodynamics.

For example, suppose an innovative astrophysicist, having access both to massive remote-control robot spaceships and to a nearby black hole, were to set out to refute the law of entropy increase. She could divide up the world categorically into high-entropy and low-entropy segments; disordered gases would fall into the former category, for instance, and ordered crystals would fall into the latter. She could place boxes of melted snow in an isolated site clearly separated from the one where she would put intricately carved ice sculptures. Finally, she could use her robot spaceships to pick up all of the high-entropy elements and toss them into the black hole, leaving only the low-entropy objects. Hence, a world of much lower entropy would ensue and the astrophysicist could legitimately claim to have foiled the second law.

The only way to avert this perplexing situation, scientists reasoned, was to suppose that the black hole itself possessed the property of entropy. In this view, the high entropy of material falling into the black hole would then be imagined as having been absorbed into the black hole's structure. Thus, the overall entropy of the universe needn't decrease because of this engulfment and the second law of thermodynamics wouldn't necessarily be violated.

But what observable property of a black hole would physically constitute its entropy? Recall that, according to the "no hair theorem," the only basic properties of a black hole that could be determined are its mass, charge, and angular momentum. To this list might be added a few more related secondary features: its volume, radius, and surface area. From the set of these quantities, all other physical characteristics must be derived—including, of course, entropy.

Scientists searched hard for the mysterious connection between the black hole entropy and its known properties. Finally, the answer was found by a Princeton University research student, Jacob Bekenstein, who obtained this result by using some of the earlier work of Stephen Hawking— namely, his result that the surface area of a black hole can never decrease. Bekenstein made the bold step of equating the black hole's entropy with its total surface area. By assuming this simple equation, he showed how the entropy of a black hole, like its area, could never diminish. This principle is known as the *second law of black hole thermodynamics*.

At first Hawking doubted that Bekenstein's postulate was valid. The reason for this concerns an alternative, equivalent statement of the law of entropy: heat always passes from hotter objects to colder objects. Black holes, by possessing entropy and energy, must also have characteristic temperatures. And naturally these temperatures must be greater than that of deep space: absolute zero. Therefore, according to this line of reasoning, because black holes are hotter than their surroundings (space), heat radiation and particles must be given off by them. But black holes can't *give off* heat energy or other forms of light, Hawking argued at that time; they're pitch black!

Then Hawking performed some calculations of his own and obtained the same amazing result that black holes are light emitters and are not really black after all. He officially declared a change of heart on this matter and joined forces with Bekenstein in asserting that black holes have entropy and temperature. George Greenstein, a professor of astronomy at Amherst College and author of a book on the subject, recalls seeing Hawking announce his findings to the scientific community in 1976 at a talk entitled "Black Holes Are White Hot":

> The talk he gave that day I will never forget. Its effect on me was electric. On the other hand, it was also mystifying. The discovery he was reporting was so remarkable, so revolutionary and so unexpected, that I had no way of assimilating it. Nothing Hawking said could be fitted into the picture of black holes that I had grown used to. . . . It seems clear that Hawking's discovery will rank as one of the great scientific achievements of our lifetime.

Hawking's proof that black holes emit energy, and hence decay, involves the use of quantum mechanics. *Quantum theory* asserts that empty space is not purely an empty vacuum but rather contains a virtually infinite quantity of paired particles and antiparticles (oppositely charged particles) hidden within its surface. It turns out that the gravitational fields of black holes at their boundaries (event horizons) are strong enough to wrest nearby antiparticles from their particle companions. The result is that a steady stream of antiparticles is continuously drawn into the surface of the collapsed star, leaving behind an enormous quantity of particles. These left-over subatomic objects disperse into space, creating the illusion that the interior of the black hole is actually radiating. It is, in fact, the immediate exterior of the black hole that produces the energy and matter drain. Thus, in summary, it seems that black holes must evaporate because of the particle-antiparticle pairs created and separated near their event horizons.

Because of this steady loss of energy and mass, black holes clearly are impermanent and, contrary to what was generally believed before Hawking's discovery, do eventually decay away to nothing. However, average-sized black holes formed by ordinary stellar collapse can be considered, for all practical purposes, to be eternal. On the other hand, in the case of ultratiny primordial black holes formed during the turbulence of the early universe, their decay is supposed to have taken place relatively rapidly. Thus, it is thought that few of these objects have survived. In general, the longevity of a black hole is directly proportional to its mass: hefty black holes live much longer than puny ones.

It has been said that all things must pass. Before Hawking's astonishing breakthrough, this maxim did not seem to apply to black holes. Now, however, we are humbled by the fact that all objects, even fortresslike gravitational strongholds with an immense capacity for self-enclosure, must yield up their controlling power, little by little, until all of their energy has dissipated into nothingness. Even black holes, symbols of dark, boundless eternity, must eventually succumb to a higher force: the will of nature. Natural decay, it seems, is a power stronger than even gravity.

Black holes, first merely thought of as extreme theoretical examples of the seemingly unlimited malleability of space-time, have thus proved to be more than just pure mathematical abstractions. Scientific evidence has shown that their explosive births and placid deaths are very real occurrences that are strongly related to the actual collapse of massive stars. And Hawking has brilliantly shown that, in between these initial and terminal events, black holes lead lives of quiet expiration—slowly exuding captured energy and material. Furthermore, detailed measurements taken from X-ray probes of binary star systems have provided growing experimental support for the full acceptance of black holes as highly tangible and very common astrophysical objects.

There is a limit to the information that can be obtained about black holes purely by viewing them from the outside. Without direct exploration of their perpetually masked interiors much of the discussion about the dynamics of black holes will forever remain speculative. To probe these dark expanses, however, would be completely futile if the probers could not emerge to tell the tale, or if they couldn't even beam their information out from these interiors back to earth. For such exploratory missions to succeed, some way would need to be found for travelers to flee the event horizons of black holes and return intact to earth.

Interestingly, the Schwarzschild and Kerr black hole models do provide much hope for connections between different parts of the universe—which could conceivably serve as escape routes for explorers. The space-time geometries of these models are so complex that they theoretically allow for tunnels through their interiors linking otherwise isolated areas of the cosmos (or, alternatively, of two *different* universes). It is fascinating to ponder the question of whether or not these tunnels could serve as ways of fleeing from black holes after being entrapped by them.

Moreover, if black hole tunnels were indeed traversable, could they serve as star gates of the sort described in *2001: A Space Odyssey*? Could they thereby provide potential means for rapid interstellar transport? Might then a series of black hole linkups revolutionize transgalactic commerce and take our earth out of its isolation in the cosmos?

There are two answers to this set of questions: one theoretical, the other practical. Sadly, the theoretical prospects for safe and speedy interstellar transportation via black hole tunnels appear to be virtually nil at this point. Too many technical problems (to be discussed in the next few chapters) with the implementation of such a program have been found. Although the astrophysical theory of black holes tempts us with exciting notions of cosmic gateways, it also taunts us with frightening prospects of perishing in attempts to use them.

Experimentally, however, the only way to determine whether or not black hole escape routes and interstellar gateways exist would be to plunge in and "test the waters." There is a distinct, though small, chance that certain elements of astrophysical theory are wrong and that black holes are indeed escapable; only manned missions could supply the proof one way or the other. It is unclear, though, why anyone would want to take such an enormous risk. For almost certainly the experience of falling into a black hole would be a nightmare beyond all nightmares.

It is fascinating—and terrifying—to imagine what such a journey to a black hole might actually be like.

*T*aking the Plunge

Floyd G. Nevish is about to embark on a special mission, one for which he has been carefully prepared for the past few years. As the first man to explore the interior of a black hole, he has been accorded all due honor and respect—and has been told that he will receive a presidential medal and considerable cash if he succeeds in returning. His trip is sponsored by the Jovian Industries supply company, a spaceship firm that hauls raw material, including uranium, platinum, gold, and other rare minerals, back to earth from Jupiter, Saturn, and other outlying planets. In addition, it carries finished products, food, and other supplies out to its many colonies spread out over the outer solar system. Jovian Industries hopes to expand its base of operations to other stellar systems; unfortunately, the lengths of such journeys would be far too great to make them worthwhile. Customers certainly couldn't afford to wait literally thousands of years for interstellar cargo ships to return. When a black hole was found about a trillion miles away from the sun, the executives at Jovian rejoiced. Reputable physicists working for their firm had advised them that some theories predict that such objects could be used as shortcuts across space. Although many of them had also warned of possible dire consequences for black hole entry, Jovian simply ignored the naysayers and decided to send their key explorer, Floyd Nevish, into the heart of the maelstrom to take careful measurements and to search for an escape route into another part of the galaxy.

Floyd has been put into cryonic suspension for the bulk of his long journey to the black hole. Shipboard computers have been programmed to wake him right before he reaches the collapsed star. As the ship blasts off into space from its Saturn orbit, the sleeping astronaut sinks into a curious dream: he imagines himself as Alice falling down an immense, dark chasm into Wonderland. Hours, days, and months pass for the dormant explorer, while enormous rockets propel the vessel toward its remote target.

Suddenly Floyd awakes to flashing lights, roaring sirens, and blinking computer screens. An electronic voice switches on and bellows, "We are now approaching the boundary region of a medium-sized nonrotating black hole of Schwarzschild type. The black hole radius is about two thousand kilometers, and its mass is over seven hundred times that of the sun. Our ship is now traveling at 99 percent of the speed of light toward the center of the collapsar."

Floyd stares nervously at a nearby screen, which depicts a black disc surrounded by a ghostly halo of radiation. Meanwhile, the gravitational forces begin to tug at his ship; invisible powers start to pull it into the waiting clutches of the dark "creature" ahead. He starts to feel some strain on his neck and torso but at first attributes these uncomfortable feelings to stress. Gradually, these forces begin to become painful and start to stretch his legs a bit. He thinks of the passage in *Alice in Wonderland* about the cake that, when eaten, causes Alice to spring up toward the ceiling, stretched out like a giant. Although at first he hopes that the sturdy makeup of his vessel will help it to withstand the tremendous pressures that are starting to affect it, he soon begins to panic and switches on the distress beacon. An emergency signal is immediately sent out by the rapidly falling craft. But this is to no avail, since, unbeknownst to Floyd, the ship has already dropped below the one-way event horizon; thus, backward communication has become impossible for him.

Floyd has not noticed the passage through the event horizon for good reason: when an object travels through such an invisible membrane absolutely nothing physically happens to it. No "bump" indicates its presence, and no dramatic change in the structure of space provides a clue as to its whereabouts. Hence, no devices are able to detect it. Nevertheless, once a ship has journeyed through it, no objects, including light signals, could ever be sent from the vessel back to the outside world, nor, of course, could the ship itself escape. For Floyd, passage through this one-way membrane has meant that there is no way of obtaining help from his sponsors.

What would the executives at Jovian Industries see if they could watch Floyd as he was entering the horizon? First they would notice time start to decelerate for Floyd: he and his ship would appear to move more and more slowly. Floyd's gestures would seem more and more sluggish, as if he had been filmed by slow-motion photography. Meanwhile all clocks aboard the ship would seem to tick at a more and more delayed pace. Finally, just before his vessel seemed about to pass through the membrane, he would appear to stop moving altogether, his visage frozen onto the surface of the horizon like a fly stuck to flypaper. Thus, to the executives, it would seem as if Floyd never actually fell into the black hole but merely stopped moving once he reached the horizon. Because no indication of motion would ever appear to anyone in the outside world, viewers might conclude that Floyd's trip was infinite and that he could never reach the interior of the black hole.

Nevertheless, Floyd *does* reach the inside of the black hole in a finite amount of time; to him, his ship's clocks appear to move at a normal pace as he passes through the horizon. Time flows inexorably forward for him, in spite of the fact that, if clocks and rulers external to the black hole could measure the coordinates of his motion, space and time would seem to be reversed for him. It is most unfortunate for Floyd that his personal clock does continue to tick, because it means that he cannot realistically hope to delay the inevitable descent into the interior of the collapsed star. He cannot turn the ship around; he is doomed to be drawn closer and closer to the crushing center of the invisible maelstrom.

Nor can he prevent the black hole's enormous gravitational forces from stretching him out and pulling him apart, while squeezing the sides of his body more and more. He is falling into the center feet first; therefore, the forces on his feet are much less than those on his head. The resulting difference in the strength of gravity on various parts of his body causes him to be tugged at different rates in various directions. Meanwhile, the actions of other components of the gravitational forces cause him to be squeezed out like a tube of toothpaste. These enormous tidal forces, unchecked by the relatively weak supporting structure of the spaceship, stretch and squeeze the vessel as well—until its framework is strained to its limits.

Floyd thinks once more about the scene from *Alice*. Then, his mind turns to an even more frightening image from his childhood: that of a candymaker pinching and pulling a rather gooey piece of turkish taffy. He remembers how the confectioner used to stretch out the sweet with his hands until it became longer and longer, while pushing its sides in with his fingers until it became thinner and thinner. Then he would crunch up the stringy concoction into a ball, pull it out once more, and start the same process over again.

Meanwhile, Floyd's tired brain prompts him to look at the computer screens one last time. All readings for external pressure and temperature are way off the scale; the spacecraft is being pulverized by the increasingly powerful tidal forces of gravity. Soon, he notices that his bones are beginning to crack; the black hole is literally beginning to rip his body into pieces. Just before the ship's final destruction occurs, he is rendered unconscious, mercifully spared the gruesome details of the catastrophic last minutes of the voyage.

What caused this flawed mission to fail so badly, one might wonder. Why didn't Floyd have time to search for an escape route? The reason for the sudden and explosive destruction of Mr. Nevish is simple. For a *nonrotating* (Schwarzschild) black hole of several hundred solar masses, or even several thousand solar masses, the gravitational strength at any point on or near the event horizon would be more than enough to destroy all but the sturdiest of objects. It would be impossible to enter these average-sized Schwarzschild objects, ranging from about ten to ten thousand solar masses, without being stretched out in the direction of motion and compressed in all other directions; while one's length increased tremendously, one's volume would diminish to zero.

According to theory, the amount of acceleration produced by the gravitational fields of Schwarzschild black holes on objects passing through their event horizons is inversely proportional to the squares of the black hole masses and directly proportional to the lengths of the approaching objects. For a "light" black hole of ten solar masses, this means that a five-foot human being passing through its horizon would experience an acceleration of over 500 million feet per second per second. This is far greater than any acceleration ever endured by anyone in the history of humankind and would certainly result in instantaneous death for the traveler.

A person entering a black hole of even *one thousand* times the mass of the sun would not fare much better. At the event horizon of such a compressed object, the acceleration of such a voyager would still be greater than fifty thousand feet per second per second, a rate much higher than that ever experienced by any astronaut in space—and virtually certain to cause immediate death.

Only for black holes of ten thousand or more solar masses might survival be possible for an approaching astronaut. Then, the rate of velocity increase would be reduced to a "mere" five hundred feet per second per second or less, a value much greater than typical accelerations on earth, but nonetheless still possibly endurable by hardy explorers. Unfortunately, however, it is extremely unlikely that such colossal objects would be located in our part of the galaxy; more likely than not, they'd be found only in the thick central region of the Milky Way, far away from planet earth.

Floyd has not really died, it turns out. After the "spaceship" door opens and a team of company doctors from Jovian Industries enter the chamber and revive him, Floyd realizes that he has survived a rather turbulent ride in an extremely sophisticated flight simulator. "We wanted to see how you would handle the stress of a black hole journey," he is told. "Of course the *real* black hole that you'll be exploring will be much, much larger. Only an idiot would attempt to voyage into a black hole of only a thousand solar masses. It would mean quick and certain death. Nevertheless, a good astronaut should be prepared for even the most extreme of circumstances."

Floyd faints. When he comes to, he realizes that he is back in a similar space vessel, soaring at a mind-boggling rate through the deep interstellar void. Could this be the real mission? he wonders, as he hears the computer monitors click on. He experiences a dreadful feeling of déjà vu as he listens to an electronic voice announce a message:

"This is the shipboard computer. We have just entered the event horizon of a large Schwarzschild-type black hole of about 10 million solar masses. We are due to reach the central singularity of the collapsar in approximately one minute. I suggest that you begin to construct aversion strategies. Have a nice day!"

Floyd breathes a sigh of relief as he looks at the ship's external pressure and temperature monitors. There is absolutely no sign that the vessel is being strained by tidal forces. Apparently, the large size of the black hole has made it possible for the net acceleration of the vehicle, as it travels just within the event horizon, to be fairly low. Floyd's arms and legs can move freely; there is no indication of discomfort as yet.

Suddenly, the space vessel is hit by a burst of high-frequency radiation. This is followed by another blast of deadly gamma rays. As the ship nears the center of the black hole, the space vessel's increasing proximity to the crushing singularity there means that it must pass through a series of bands of intense radiation given off by decaying matter. Floyd quickly realizes that if he doesn't soon manage to escape from the black hole, he will, in a short while, be exposed to a lethal dose of radiation emanating from the crushed mass at the very core of the collapsed star. Panic-stricken, he hastily searches for a way out of his dire predicament.

All of a sudden, almost like the Cheshire Cat's smile in

Lewis Carroll's story, a gap in the firmament appears out of nowhere. Elated, Floyd immediately concludes that this aperture must be the long-sought entranceway to a Schwarzschild tunnel that leads to another part of the cosmos. When his spaceship falls into the opening of the tunnel, he braces himself for a quick escape out of the black hole. Much to his horror, though, once he is inside the throat, it immediately begins to collapse in on itself. This is detected by a gravitational wave device; the throat can't actually be seen. Nevertheless, the effect is terrifying. Floyd cries out in anguish as the walls of the tunnel move closer and closer to the spaceship. Exhausted, he faints once again as his ship is engulfed by a vast nothingness.

It seems that large Schwarzschild black holes, though navigable inside, would not permit ready escape from their interiors and might, in effect, prove just as deadly as smaller ones. Although a black hole of 10 million solar masses would have a distance, from event horizon to singularity, of more than 15 million miles—encompassing a comfortably large volume, over thirty thousand times the size of the sun, in which to travel—the freedom of motion in this region would be severely hampered by strong gravitational forces. A space traveler would have little choice but to endure a dreadful wait while gravity tugged his ship closer and closer to the central domain of infinite curvature. Meanwhile, deadly radiation, caused by the destruction of matter within the interior, would almost certainly kill the space traveler as he approached the singularity.

The escape tunnels predicted by Einstein's theory of relativity would, in the case of the Schwarzschild metric, most likely prove to be extremely difficult to cross. These passageways, sometimes called *Einstein-Rosen bridges*, would exist for only a brief period; hence the mouths of such gateways would be dynamic, not static. Like camera shutters, these apertures would quickly open, then close again, trapping all objects unlucky enough to have entered within their throats. This expansion and recontraction would be so rapid that, unless one moved faster than the speed of light, one would undoubtedly be caught by the recontraction and crushed by the enormous tidal forces of gravity.

Gateways for which passage would be possible only by traveling at faster-than-light speeds are called *spacelike connections*. Plotted on a space-time diagram, a line segment depicting the displacement between the two end points of these tunnels would appear to tilt far more along the spatial axis than along the temporal axis. Technically speaking, that means that the metric signature could be said to be negative (see Chapter 2). Consequently, unless the faster-than-light restriction of special relativity could be circumvented, there could be no causal relationship between the two ends; in other words, neither signals nor spacecraft could be sent between them. Hence, Schwarzschild tunnels would be impassable except at faster-than-light speeds.

Kerr Tunnels

Kerr-type, or rotating, black holes would represent far more intriguing prospects for interstellar shortcuts than Schwarzschild models. Nevertheless, Kerr black holes would have serious traversability drawbacks as well—ones peculiar to rapidly spinning collapsed objects. Let's send our brave fictional astronaut, Floyd G. Nevish, back into space one last time for a close-up look at the complex properties of Kerr-type objects.

Using highly realistic flight simulators, Floyd has now completed his two black hole training missions, experiencing firsthand the lethal conditions found in two different sorts of Schwarzschild objects. He is now fully prepared for his true test of courage: to board an actual spaceship and embark on a perilous mission to a massive Kerr-type black hole. The collapsed object chosen for him to explore is a rotating structure of 10 million solar masses having a radius of about 20 million miles. This massive object is so far away that Floyd must be placed in deep freeze for several years, waking before then only in the case of emergency. During his flight literally hundreds of years will pass on earth, an outcome of the special relativistic prediction of significant discrepancies between the rates of clocks located in different frames of reference.

Floyd is now placed in the hibernation unit. After his two apparent brushes with death, in his previous "voyages" to Schwarzschild black holes, he is understandably quite nervous that something might go wrong with his mission. On being assured that Kerr-type objects, because of their rotation, are believed to have radically different properties and have a far greater chance of being escapable than Schwarzschild holes, Floyd dozes off to a deep, restful sleep.

He wakes up to the sound of a rather reassuring electronic voice, similar to ones he had heard on his previous missions:

"Good morning, Commander Nevish. I wish to inform you that we are now just under 20 million kilometers from the boundary region of the Kerr object. Traveling at our current rate of 99.9 percent of the speed of light, we should reach the border zone in less than one minute."

Glancing at the viewscreen, Floyd observes a dark circular expanse directly ahead, blotting out most of the stars in the sky and distorting the light paths of many of the still visible ones. Then he turns to his vessel's control unit and meticulously examines its detailed readings. Following careful instructions received earlier, he makes sure that the spaceship is heading directly into one of the poles of the rotating black hole, and not into the equatorial region.

He recalls from his training that there would be a tremendous difference between entering a Kerr-type black hole via its axial or equatorial directions. Because of peculiarities of its spinning structure, approaching it via its equator would almost certainly lead to a crushing demise; along this band there would likely be infinite gravitational tidal forces. These tidal forces would not, however, be of any great strength or significance near the poles of the black hole. Therefore, Floyd resolutely decides to steer clear of the equatorial region by using his computer's guidance system to make careful adjustments to the ship's trajectory. Finally, when he is absolutely sure that his craft is bound for the relatively safe "north pole" of the black hole, he braces himself for passage through the boundary of the dark vortex. He places himself in a special device designed to reduce the impact of extreme gravitational turbulence.

Floyd's precautions are completely unnecessary; there are

absolutely no signs of gravitational stress or turbulence when the ship passes through the event horizon of the collapsed star. This absence of tidal forces is largely due to two factors: the enormous size of the rotating black hole and the entry of the ship via one of the object's poles instead of through its equator. Relieved that the mission is proceeding as planned, Floyd begins to plot out a safe path through the black hole's dark interior.

In contrast to the regions lying near the centers of Schwarzschild black holes, it has been theorized that the interiors of Kerr objects would be readily navigable by spaceship; that is, it would be possible to steer a ship along a trajectory that avoided exposure to the intense tidal forces that lie near the center. This important distinction stems from a difference in the shape of these objects' singularities: whereas the static Schwarzschild model has a *point* singularity, the rotating Kerr model contains a *ring* singularity.

Ring singularities would be, according to theory, just as deadly if approached as point singularities. Like their dimensionless cousins, they would embody strange domains of infinite space-time curvature—crushing nearby objects with their enormous tidal forces and poisoning them with their tremendous quantities of radiation. Nevertheless, the hollow structures of such zones would enable a traveler to avoid them indefinitely merely by carefully steering all spacecraft through the nonsingular centers of these rings. Thus, for the space traveler, they would offer much more hope of long-term survival.

Floyd is in an optimistic mood as he detects the ring singularity on the screen and plots a steady path directly through its nonsingular middle region and right into the throat of the Kerr tunnel. The ship lurches forward and easily reaches the central passageway in a short time. As expected, the craft begins to move smoothly between the relatively narrow walls of the throat.

However, within a matter of seconds, Floyd's good fortune proves to be short-lived. The Kerr tunnel that surrounds his ship has started to collapse in on itself. In a terrifying avalanche of decay, crushing forces have begun to smother his ship. Meanwhile, a tide of radiation has started to engulf his spacecraft in a lethal bath of energy. Floyd prays for some way out of this dilemma, but alas, he cannot escape. Unfortunately for him, unlike his previous trips to black holes, this is not just a simulation.

Floyd's fatal error was to assume that Kerr tunnels would be permanent structures. Indeed, Kerr tunnels by themselves would be highly stable, in marked contrast to Schwarzschild gateways, which would automatically collapse after brief intervals of existence. Nevertheless, Kerr tunnels would still be highly sensitive to outside perturbations; the slightest disturbance would lead to structural collapse. Even the presence of average-sized spacecraft would be enough to send these tunnels tumbling in on themselves, immediately crushing everything within. These gravitational forces would naturally be accommodated by lethal bursts of radiation, stemming mainly from the rapid decay of the captured material.

It is possible, however, that Kerr tunnels could be stabilized against collapse. An advanced civilization, with access to powerful new materials and forces, might somehow develop the technology to construct vast bulwarks against decay of these gateways. Citizens of that society could then explore these passageways without fear of imminent destruction—and, unlike our fictional protagonist, could emerge from them completely unscathed. Thus, it is conceivable that passage through Kerr tunnels will someday be safe for astronauts.

A Trip through Nomad's Land

What would an astronaut see after passing successfully through the stabilized throat of a Kerr tunnel? One distinct possibility is that the explorer would emerge from an identical copy of the original collapsed star, albeit in another part of the cosmos. The most widely accepted theoretical interpretation of Kerr-type tunnels has been that they would link two large black holes, located in two otherwise separate regions of the universe. According to this line of reasoning, by journeying through such a gateway, an astronaut could be instantaneously transported to another star system hundreds of trillions of miles away.

This is also the popular media portrayal of these tunnels, championed in such movies as *The Black Hole* as well as in numerous science fiction stories about transgalactic journeys through black hole gateways. According to these exciting epics, based half on fact and half on fantasy, voyaging through a black hole gateway would be as simple as driving through the Lincoln Tunnel (during off-peak hours, that is). Naturally, there is no mention in these narratives of the many complications that might occur during black hole passages.

Unfortunately, "real life" Kerr-type connections between two black holes would, most likely, be far more problematic. Suppose an explorer did manage to travel from one black hole to another via a Kerr gateway. What could he do then to escape back to the world outside the second collapsed star? Absolutely nothing. He would be stuck within the event horizon of this duplicate black hole; clearly, he couldn't realistically hope to pass intact through the one-way membrane of the collapsar. Instead, he would be painfully reminded of the fact that an event horizon is essentially an enormous invisible prison.

Therefore, this unfortunate astronaut would be forced to reenter the Kerr tunnel and try his luck at escaping through its other terminus. Sadly, it is likely that he would be faced with the same obstacle; he wouldn't be able to traverse the one-way membrane of the other black hole. All he could do there would be to turn around again and travel through the gateway once more.

Thus, it seems that a gateway between two Kerr black hole terminals, instead of acting as a rapid means of interstellar transport, would represent a closed system with no possible means of escape. Astronauts unlucky enough to enter one of these tunnels, assuming that it was stable, would be stuck in a virtual "nomad's land." Like desert nomads or gypsy vagabonds, they would be compelled to wander forever, unable ever to return to the real, physical universe. It would be as if they were stuck in a bizarre subway system with all entrance turnstiles and no exit turnstiles—nor any other means of leaving the network.

The doomed adventurers would rue the day that they ever allowed themselves to be captured by a black hole. Their aimless wanderings would likely persist until they ran out of provisions completely and were forced to abandon all hope. Truly, they would be among the loneliest and most desperate individuals in the cosmos.

Emergence into a duplicate black hole is just one of the possible scenarios for passage out of Kerr-type tunnels. A second, more hopeful, possibility involves departure via a completely different type of object altogether—a sort of inverted black hole, called a *white hole*. Let's now examine the intriguing prospects for the formation and utilization of gateways between black holes and white holes.

CHAPTER 4

WHITE HOLES: OUT OF THE LIGHT

Cosmic Gushers

A magician pulls a rabbit out of a hat. He reaches his hand in again and removes a large chicken. This is followed by a pair of doves. A black baton is pulled out next. Then comes a red silk scarf, so long that it seems to take forever to remove. Amazing the audience once more, he pulls out an even longer blue scarf, tied to a purple one with pink polka dots.

The idea of getting something from nothing is so alien to our experience that we are fascinated and befuddled to see wondrous things appear out of the blue on the stage. These shows are entertaining partly because of our shared gut feeling that conjuring cannot really occur and that, in truth, there is "nothing new under the sun."

Maybe not under our sun. But out in space, far away from familiar stars and planets, there may exist fabulous celestial objects called *white holes* that spew out matter and energy sporadically, gushing up everything from neutrons to newts and from photons to futons. With their extraordinary ability to conjure up a virtually unlimited range of material objects and energy forms and to cause these substances seemingly to arrive from nowhere at all, white holes are magicians on a cosmic scale.

According to theory, whatever a black hole might devour, a white hole could spit out. If it's possible for a black hole to consume a Mercedes-Benz, a white hole could certainly produce an identical car. Yet, since most of what a black hole gobbled would be in the form of cosmic radiation and interstellar dust, it is very likely that most of what a white hole disgorged would be in the form of radiation and dust as well.

The reason for this strange symmetry is related to the origins of white hole theory. The idea of white holes stems from the early general relativistic idea that the Schwarzschild solution, possessing a future singularity, must possess a past singularity as well. If one maps out the full spacetime diagrammatic depiction of a Schwarzschild object, one finds that a section containing a past singularity is a natural extension of the model. Whereas the future singularity is in the top, or "forward time," half of the diagram, the past singularity is in the bottom, or "backward time," half of the diagram. If one identifies the upper part of the picture with the all-consuming singularity at the center of a black hole, one may also assign the lower part to the ever-gushing singularity at the center of a white hole. Thus, white holes, according to this approach, are the time-reversed images of black holes. Whatever a black does, a white hole does the time-reversed opposite.

One might think of this phenomenon in terms of motion pictures. Imagine a movie taken of an automobile sinking down into a massive pool of quicksand. Now suppose that this same movie were shown in reverse; the car would appear to move upward and emerge intact from the mire. Similarly, a backward version of a film showing an object falling into a black hole singularity would depict the same object emerging from a white hole singularity.

Time-reversed singularities are not just found in the Schwarzschild solution of Einstein's equations. The Kerr solution, containing future "black hole–type" singularities, embodies singularities of the past as well. It is clear that a full accounting of gravitational phenomena must include reference to white objects as well as to black ones.

Where might time-reversed singularities be found? One obvious choice involves the ultimate white hole: the "big bang" origins of the universe. One of the experimentally verified predictions of Einstein's theory of general relativity is that the universe is expanding; its galactic clusters are moving farther and farther apart from each other. Tracing this expansion to its source, scientists have concluded that the physical universe was formed in a fiery big bang some 10 billion to 20 billion years ago. All matter and energy present today can be said to have originated in this sudden explosion.

It is remarkable how much a time-reversed film of the big bang eruption would look like the instantaneous gravitational collapse of a fire ball. Or, conversely, how much a backward-in-time image of the decomposition of energy into a black hole singularity would look like a forward-in-time picture of the creation of energy from a big bang explosion. It is no wonder that most physicists refer to the creation of the universe as the *initial singularity*. From this singularity of the past, it is believed that the entire energy and matter content of the cosmos gushed out in a single burst of power.

Why only *one* burst, though? Why haven't there been many such explosions in the past? Clearly, during the past 10 to 20 billion years, there couldn't have been more than one *colossal* big bang; there is absolutely no evidence for additional explosions of universal magnitude. However, in the history of the cosmos, it is quite possible that there have been a series of "little bangs": explosions involving small parts of the universe. As in the case of the "big bang" itself, these relatively tiny fonts of energy would represent true singularities of the past: rapidly gushing time-inverted versions of black holes. In other words, they would be white holes.

In 1965, the Soviet scientist Igor Novikov and the Israeli physicist (and future science minister) Yuval Ne'eman independently developed the first comprehensive theories of white holes, which Novikov referred to as "lagging cores." The basis of their theoretical work is the idea that the creation of the universe may have been a multistage process. According to this viewpoint, most of the cosmos emerged from the initial big bang explosion, but sizable chunks of energy and matter have continued, over time, to emanate from lagging regions of the blast.

One might wonder how such a delay would have come about: what would have prevented the universe from being created all at once? Nothing, in fact, would have compelled delays in parts of creation, but nothing would have precluded them either. Time lags in the emergence of the physical universe may or may not have actually happened; Novikov and Ne'eman merely urged other theorists to keep an open mind about the chance that "little bangs" may have occurred after the initial big bang.

Several years after Novikov and Ne'eman proposed their "lagging core" theories, astronomers began an extensive search for these objects. They scanned the sky carefully for signs of celestial bodies that emitted large quantities of intense radiation. They particularly sought out objects that produced light energy in sporadic doses, thinking that this would provide greater evidence that they were cosmic gushers. After John Wheeler coined the term *black hole*, the expression "lagging core" was dropped from astrophysical parlance and replaced with the moniker used today; ironically, the term *white hole* is a far more colorful way of describing these gushers. The sightings, in the mid-1970s, of black hole candidates such as Cygnus X-1 only continued to spur on the intensive search for their seemingly time-reversed opposites.

The strongest white hole contenders during the 1970s were the extremely distant, extraordinarily bright celestial bodies known as quasistellar objects (*quasars*). Quasars were discovered at the beginning of the 1960s through data obtained by radio telescopes. It was found that they emitted continuous streams of intense radiation, far more than that produced by any known celestial bodies of comparable size. One possible explanation, advanced by physicists such as Novikov, for a quasar's source of power was that it contained a central white hole singularity, emitting a never-ending torrent of light energy. However, detailed theoretical work during the late 1970s proved that white holes would be too unstable to drive quasars. Other explanations for quasars were soon found, removing much of the impetus for white hole investigations. Nevertheless, even today many astrophysicists continue to claim that they have discovered viable white hole candidates. It is still not at all clear whether these objects are real physical entities or mere mathematical abstractions.

Stepping Out

If white holes actually do exist in the physical universe, they would provide strong hope for the possibility of escape from Kerr tunnels. Clearly, if the gateways present in Kerr-type black holes were to connect only to other black holes, using these tunnels for interstellar travel would be highly problematic—an unlucky explorer would find himself trapped forever.

However, an alternate explanation for Kerr tunnels has been proposed, according to which, instead of joining two black holes, the tunnels would connect a black hole in one part of the cosmos with a white hole in another part. This theory is derived from the peculiar structure of the Kruskal diagrammatic depiction of a Kerr model, a model which clearly contains both past and future singularities as well as traversable boundaries between different space-time regions (see Chapter 2). The idea is that an object could readily pass from one section of the universe containing a future singularity, namely, a black hole, to another section encompassing a past singularity, namely, a white hole. Whereas the former region of the cosmos would be surrounded by an all-enclosing event horizon, the latter would be bordered by an all-excluding "antihorizon." Comprising a sort of past event horizon, an *antihorizon* is a surface out of which things can emerge but into which nothing can enter.

Imagine, then, if an astronaut were to enter a Kerr black hole, pass through its tunnel, and then happen to surface into the interior of a massive white hole. That explorer would immediately find himself being pushed *outward* by the white hole and soon would be completely expelled from it through its antihorizon. At that point, he would likely be in a completely different region of the universe, perhaps hundreds of trillions of miles away from his home planet. In the event that he were to try, at that point, to return to his original part of the cosmos by turning around and reentering the white hole, he would be forcibly prevented from doing so: the antihorizon would block all attempts to reenter. Once he had stepped out, he would have stepped out forever.

Horizons and antihorizons are curious opposites, like entrance and exit turnstiles or sources and mouths of rivers. Horizons confine everything that happens to be inside them; antihorizons "imprison" everything that happens to be outside them. In the latter case, these prisons "surround" the remainder of the universe (aside from the white hole); they let everything inside the white hole out but prevent anything else from coming in.

It is fascinating to consider the possibility that white holes might provide escape routes into other parts of the universe. Yet unfortunately, the notion of using white holes for the rapid transport of long-distance space travelers is fraught with many apparently insoluble difficulties. One of the most troubling problems concerns the sheer unpredictability of white holes. In all of our arguments for the use of white holes as escape routes, we have assumed that they would output matter and energy in a regular and orderly fashion. However, we cannot automatically suppose that they would spew out intact whatever their black hole companions have consumed. It is quite possible that white holes might, in fact, deliver objects in a completely haphazard manner. A space explorer, having passed through a Kerr tunnel, might find to his dismay that he has been expelled by a white hole only partially intact—missing his chin, let's say—or maybe even transmitted, in scattered pieces, during a different era of the universe altogether. Unquestionably, all astronauts would wish to avoid such gruesome scenarios.

Another outstanding problem with the dual-hole cosmic transport model involves the growing theoretical evidence that white holes, if formed, would quickly and irreversibly decay. They would likely last for only seconds, either rapidly converting into black holes or exploding in enormous fireballs. Consequently, any unlucky travelers attempting passage through these volatile celestial bodies would either be stuck forever within an event horizon, in the first case, or almost immediately pulverized, in the second case. Knowing all this, it is unlikely that any sane astronaut would wish to try to journey through one of these unstable gateways.

White Holes Are Shy

Why would white holes be unstable? Why would they immediately convert into black holes? The reason for this is quite bizarre: white holes are shy and cannot stand being out in the open. Well, they're not *literally* shy, of course, but, figuratively speaking, white holes like to avoid exposure to the outside world as much as possible. Consequently, once formed they seek to render themselves invisible as quickly as possible. And, because the ultimate unseen object is a black hole, white holes aspire to be black holes, a quest that is ultimately suicidal.

This strange result, which I call the "White Holes Are Shy" theorem, was first reached in 1974 by Douglas Eardley of the California Institute of Technology (Caltech), a hotbed of black hole, white hole, and wormhole study. Eardley extensively analyzed Novikov and Ne'eman's "lagging core" models, in which small regions of matter emerge from the initial singularity at a later time than their surroundings, and reached the conclusion that these solutions would be highly unstable, decaying almost immediately. The reason for this, he asserted, was the propensity of white holes for accumulating around themselves coats of matter and energy that would ultimately smother them.

Consider a typical white hole, one modeled on the Schwarzschild solution, let's say. It would consist of a central past singularity, from which energy and matter would gush out, completely surrounded by an antihorizon. Naturally, since the white hole would appear as the time-reversed twin of a Schwarzschild black hole, the white hole would be spherical as well, and the distance between its past singularity and its antihorizon would be precisely its Schwarzschild radius.

In a black hole, material and energy would be drawn into the event horizon; here, the opposite phenomenon—exclusion of matter and energy—would occur. Hence, energetic material would continue to build up outside the antihorizon of the white hole, totally unable to enter. Layer upon layer would be added to this coating, until the white hole was surrounded by a dense shield of energy. Eardley calls this the "blue sheet," since the energy present in the coating would likely be shifted by gravity toward the blue end of the light spectrum (this is a phenomenon peculiar to general relativity).

The existence of the energetic coating surrounding it would provide a perfect excuse for the white hole to become a shrinking violet. This state of withdrawal from the world would begin when the ever-growing mass of the blue sheet, following the general relativistic principle that "matter tells space to curve," caused the space-time region around the antihorizon to become sharply distorted. Eventually, space-time would grow so warped in this region that the former antihorizon would then become a true black hole event horizon and would, from that point on, prevent all material from leaving what was once a cosmic gusher. Thus, the self-revealing white hole would have transformed itself into a self-concealing black hole.

Another way of looking at this situation is to consider the light emerging from the past singularity at the center of the object. If it weren't for the sheet of material surrounding the white hole, all light rays originating in the center would escape via the antihorizon; the white hole could, in that case, be said to be an energy gusher. Instead, the gravitational field of the coating would serve to warp space-time in its vicinity and recapture all light rays emerging from the center. Consequently, these rays would be trapped in that region and could never be seen by the outside world. Since, from that moment on, no light could escape from this body, it could be said to have become a black hole.

If an outside observer were to gaze at this series of events, they would appear most peculiar. No sooner would a white hole emerge from the spatial void than it would then coyly withdraw behind a dark curtain. Depending on its mass, each white hole would disappear at a different rate. For a white hole of ten solar masses, the conversion to a black object would typically occur in less than one-thousandth of a second; a white hole of even a million times the mass of the sun would survive for only slightly more than a minute.

Thus, if white holes were created during the early phases of the universe, it is most unlikely that any would still be left; after a brief interval, all of them would have been converted into black holes. Even white holes created in later eras would be long gone—forever hidden behind event horizons.

With the chances of white hole long-term survival virtually nil, the prospects for using them as interstellar gateways are similarly bleak. Hence, most theorists have concluded that Kerr tunnels, whether linking two black holes or joining a black hole to a white hole, would not be usable as traversable star gates. Another means of interstellar transport would need to be found, one clearly based on general relativity, but specially constructed to be navigable. An alternative "traversable wormhole" scheme, currently under development at American and European universities, seems to have far more promise of success.

Before turning to the topic of wormholes, I'd like to take one last look at white holes and what physicists now have to say about these apparently extinct creations. For example, one current white hole theory contends that these objects really do exist but are entirely indistinguishable from black holes. Could white holes and black holes be one and the same?

Masquerade Party

As any native of England knows, there is an intense rivalry between Oxford and Cambridge universities, one that transcends all reason and that often seems to resonate through the very fabric of the universe itself. This friendly, but fervent, sense of competition extends from sports to science and permeates the lives of those teaching at or attending these centers of learning. It is no wonder, then, that the world's two leading experts on the theorized history of the universe, Roger Penrose of Oxford and Stephen Hawking of

Cambridge, beg to differ in many aspects of their philosophical interpretations of cosmic events.

Their conjectural descriptions of white holes are no exception to this rule. Hawking is convinced that black holes and white holes are two sides of the same coin; Penrose is equally convinced that, aside from the big bang, white holes have never actually existed. Each has strong arguments to justify his point of view, based on a fundamentally different conception of reality: Hawking believes that the overall history of the universe is essentially time-symmetric, whereas Penrose feels that it is basically time-asymmetric.

The difference between time-symmetric and time-asymmetric series of events is quite simple. In the former case, there would be absolutely no difference between a backward and forward showing of a film depicting these occurrences; in the latter case, there would be an unmistakable distinction between the backward-in-time and forward-in-time versions.

Clearly, the universe is neither exclusively time-symmetric nor completely time-asymmetric. Generally, on the microscopic scale—the world of particle interactions—events take place in a time-symmetric manner, whereas on a macroscopic scale—the world of solids, liquids, and gases—events occur asymmetrically in time. Newton's laws of motion, completely symmetric in time, apply very well to small-scale happenings such as the collision of two tiny pellets, whereas the second law of thermodynamics, asymmetric in time, applies equally well to large-scale occurrences such as the melting of ice.

Penrose and Hawking both agree that both time-symmetric and time-asymmetric elements play important roles in the physical universe. They differ, however, in their interpretations of the relative significance of each of these two sorts of relationships.

Penrose advocates the viewpoint that the law of entropy —and hence a fundamental time asymmetry—is built into the very structure of the universe. He contends that the universe must have started out in a perfectly orderly state of low entropy and must terminate someday in a completely disorderly state of high entropy. The basis of this asymmetry, he asserts, is an essential difference between the space-time structures of the beginning and end of the universe: the beginning would contain only one regular, homogeneous (perfectly similar for all points) event, the big bang itself, while the end would embody a series of irregular, inhomogeneous happenings, such as sporadic decays into black hole states. According to Penrose, the law of entropy is a result of this fundamental discrepancy between past and future singular states, stemming from the contrast between the smooth initial singularity (big bang) and the erratic final singularities (black holes).

Where might white holes fit into this scheme? Well, they wouldn't. Penrose argues that white holes are incompatible with the law of entropy increase. On the one hand, they would represent highly irregular singularities; on the other hand, they would exist mainly in the early stages of the universe. Therefore, Penrose excludes white holes on the basis that they would constitute erratic, high-entropy initial singularities—contradicting the rule that the initial singularity must be smooth and unique—in other words, low-entropy.

Hawking's philosophy is a fundamentally different approach to the law of entropy; he argues that it has nothing to do with the nature of the big bang or any other gravitational phenomena. Essentially, he asserts, the basic laws governing the universe are completely time-symmetric. The only reason why order is decaying into disorder and entropy is currently increasing is that, if this weren't the case, life couldn't have formed and we wouldn't be here. Because, in fact, we do exist, we must live in a sort of universe, or era of the universe, in which the law of entropy strictly holds true. (This argument is a good example of the use of the *anthropic principle*, to be discussed in Chapter 8.)

If the universe is essentially symmetric in time, there is no reason to exclude the possibility that certain particular gravitational phenomena, such as black holes and white holes, are time-symmetric as well. In 1976, Hawking made this very assertion. He proposed that black holes and white holes each emit light (thermal radiation from black hole surroundings, cosmic gushing from white hole antihorizons), to a certain degree, and each absorbs light (black hole event horizons, white hole "blue sheet" horizons), to some extent. Then, because black holes and white holes each absorb and emit, and have otherwise identical properties, Hawking concluded that they must be one and the same. In other words, white holes could masquerade as black holes, and black holes could pass for white holes, and there would be absolutely no way of distinguishing these celestial creations.

Both Penrose and Hawking do concur that the term *white hole* is largely obsolete, in the first case, because it describes a nonexistent object, in the second case, because it redesignates an already named one: a black hole. With that all said, the Oxford mathematician and the Cambridge physicist have elected to drop this apparently moot topic and turn their formidable mental powers to other subjects of interest.

From Oxford and Cambridge, the elite academic institutions of Britain, we now turn to one of the equally prestigious establishments of the American Ivy League. Traversing an instantaneous transatlantic gateway, we find ourselves situated at Cornell University in Ithaca, New York, home of one of the leading centers for observational astronomy.

CHAPTER 5

WORMHOLES: INTERSTELLAR PASSAGEWAYS

Wormholes as Gateways

Ithaca, New York, is an ideal place to think about black holes, gaps in space-time, and ripples in the cosmic fabric. Its hilly terrain, almost impossible to navigate during its long winters, is punctuated by hundreds of gorges, chasms, waterfalls, and cliffs. In Ithaca, the shortest distance between two points is rarely a straight line, and drivers spend years mastering all the twists and turns of the roads.

In a rustic-looking house, sitting high above one of the gorges, lives the Cornell University astronomy professor Carl Sagan. It is almost a cliché in the world of academia that there are good researchers and good teachers, and never the twain shall meet. Sagan is one of the rare exceptions to this "rule." He has won acclaim for his research and his teaching—and for his writing as well. Not only is he a Pulitzer Prize winner, he has also won the Smith Science Prize and the NASA Medal for Exceptional Scientific Achievement, as well as the Hugo and Peabody awards for his science fiction. He is the man who designed the interstellar messages of peace sent aboard the *Pioneer* and *Voyager* space probes in the hope that alien civilizations will someday learn about intelligent life on earth. And he is also the convincing orator who warned about the disastrous effects of a nuclear winter. However, most people know him as the mod-looking pro-

ducer, writer, and narrator of the late 1970s television series
Cosmos, a show that led viewers on a weekly expedition
through the vastness of space. Television watchers also remember his frequent appearances on the *Tonight Show*,
where he was often playfully spoofed by the host, Johnny
Carson. Overall, in the past decade, perhaps no man has
been identified more with the field of astronomy than Carl
Sagan.

In 1985, Sagan published a novel, *Contact*, describing a
first encounter between humans and another life-form. In
this story, five people experience a truly fantastic voyage
across the cosmos. On the first leg of their trip, they travel
through a sort of star gate, pass through a space tunnel, and
arrive in a sort of celestial no-man's-land. They then journey
past a region dubbed "Grand Central Station," a spatial junc-
tion where routes lead to thousands of different destinations
and voyagers can make interstellar connections. Finally,
they end up in orbit around the star Vega, billions and billions of miles away, a region of space that is clearly too far
away to reach by conventional space travel. Yet it takes these
characters less than an hour to complete this perilous voyage.

After their journey, two of the protagonists, Eda and Vaygay, speculate on the speed of their flight and the nature of
the space tunnel. The first idea that occurs to Eda is that the
tunnel is the connecting link between two black holes. However, she quickly rejects that idea for some of the reasons
mentioned in the previous chapters of this book, including
the fact that such a link would be unstable and would quickly
collapse into what is called a *singularity*: an impassable,
discontinuous point in space and time.

Vaygay suggests a few more reasons why they could not
have traveled through a black hole. A black hole passage
would have subjected them to enormous gravitational tidal
forces, strong enough to have stretched their vessel like taffy,
and a tremendous quantity of radiation, harsh enough to
have killed everybody on board. Finally, the amount of time,
as measured on earth, to have traveled through a black hole
tunnel would have been infinite. Thus, once having passed
through such a gateway, a traveler would never be able to
return home.

For all of these reasons, the characters in *Contact* conclude that they must have journeyed through a different sort of celestial object altogether, one previously unknown. But through what? This is exactly the question that Sagan asked himself while he was writing the novel.

Sagan wanted to make events in his novel conform to the laws of gravitational physics. Yet, while he was writing *Contact*, he could not think of a way to allow his protagonists rapid passage through the cosmos. He searched through dozens of journal articles; all of them showed that black hole/white hole travel was impossible, so, for the sake of scientific realism, this type of transportation was not an option. And none of these articles hinted at any alternative. Finally, he contacted a friend, the astrophysicist Kip Thorne of the California Institute of Technology in Pasadena, and asked him whether he knew of any postulated means for instantaneous interstellar travel that avoided the problems associated with black holes.

Kip Thorne, like Sagan, is a man of many talents: a renowned theorist, well-respected educator, and award-winning scientific writer. Born in Utah in 1940, Thorne advanced rapidly through college and graduate school, graduating from Caltech in 1962 and obtaining a Ph.D. in physics three years later from Princeton. He did his graduate work in astrophysics there with John Wheeler, someone with whom he would continue to collaborate after graduation. Together with Charles Misner, Thorne and Wheeler completed the definitive megatextbook on Einstein's theory of relativity, *Gravitation*, published in 1973.

Since 1970, Thorne has been a full professor at Caltech and has continued to study the nature of stellar collapse, in particular the theory of black hole formation. Thorne has been extensively involved in the search for and evaluation of black hole candidates, particularly Cygnus X-1. Since the mid-1970s he has had an ongoing bet with Stephen Hawking on whether this object is an authentic black hole or is just an ordinary celestial body, hidden from view. Thorne has bet that Cygnus X-1 is genuine; Hawking has wagered that it is a false candidate. If Hawking wins, Thorne will buy him a four-year subscription to the magazine *Private Eye*. If Cygnus X-1 is proved to contain a black hole, Thorne will be awarded one year of *Penthouse*. So far the bet has not been resolved, although Hawking has recently admitted that Thorne is probably right about Cygnus X-1.

Obviously, Kip Thorne is a man who enjoys a challenge. When Carl Sagan challenged him to find some way of making interstellar travel a possibility, he couldn't resist. With the assistance of two of his graduate students, Michael Morris and Ulvi Yurtsever, he produced an actual working model of a cosmic "wormhole," an interstellar link that was safe enough, wide enough, and stable enough that it could be traversed by spaceships in a short amount of time. Sagan immediately incorporated the results into his novel. After excluding the possibility of black hole travel, Sagan's characters turn to wormhole travel as a reasonable alternative explanation for their voyage.

In 1987, two years after the publication of *Contact*, Morris and Thorne decided to publish their results (Yurtsever joined them in a later work) in the *American Journal of Physics*, a periodical devoted to educational and cultural topics in physics. Their seminar paper is the *Betty Crocker Cookbook* of wormhole design. Morris and Thorne's nine conditions for traversable wormholes seem to provide a minimal list of requirements to ensure safe passage. Together, these rules dictate a necessary blueprint for the design of interstellar gateways.

The first rule is that the wormhole must be static and spherical. The former means that its shape and size can't change with time but must remain frozen forever. This prevents the outstanding problem with the Kerr tunnels and black hole/white hole connections described in Chapter 4, namely that these gateways shrink with time until they are impenetrable singularities. A good wormhole must remain open for passage forever, providing a permanent and secure means of transport.

The latter stipulation, that the wormhole be spherical, is merely a simplifying assumption. Regular, symmetric objects are, not surprisingly, much easier for physicists to deal with than irregular ones. It is a running joke among engineers that physicists always assume the simplest case even when it is not quite applicable. In one story commonly told by engineers, there is a physicist who owns a chicken factory. He decides that the efficiency of the factory needs to be improved, so he sets out to design a newer, better chicken-plucking machine. Sitting down to draw up the blueprints for his device, he mumbles to himself,

"Let's see now, we first assume that the chickens are spherical."

Spherical symmetry is a common assumption for physical models. But, as in the chicken factory story, it is not always the most valid one!

The second of Morris and Thorne's conditions is that the cosmic tunnel must obey Einstein's equations of general relativity. As we have seen in Chapter 2, this means that the geometry of the wormhole must be related to its material composition in the manner prescribed by Einstein. This condition ensures that the laws of relativistic physics are being followed.

How might the wormhole appear? The third rule describes just that: according to Morris and Thorne, the wormhole must look like a flattened hourglass, with two "basins" connected by a narrow "throat." Like sand passing through an hourglass from the top half to the bottom, space voyagers would pass through the throat from one section of the cosmos to another. This is, of course, only a rough analogy; the real hourglass must somehow be carved out of the very structure of space-time itself.

In this scheme, the top basin of the hourglass-shaped wormhole corresponds to our region of the universe, the bottom basin to another. These sections of the hourglass must be flat, because, as we recall from Chapter 2, matterless or low-matter regions of the cosmos correspond in general relativity to flat space-times. Clearly, in empty space, away from the center of the wormhole, there would be extremely little mass. Therefore, there must be little curvature as well.

The throat of the wormhole, a narrow piece of extremely curved space-time connecting the two flat basins, must, on the other hand, be produced by a great deal of concentrated mass. All paths in the wormhole region must start off on one basin (representing one section of the cosmos), be straight until they near the throat, then sharply curve down through the throat and arrive onto the other basin (corresponding to another part of the cosmos). Thus, Morris and Thorne's scheme precisely defines the shape of the wormhole in a manner that prescribes how vessels travel through its throat.

The fourth rule states that there cannot be event horizons. By precluding all possibility of escape, event horizons would render cosmic connections useless to travelers. Who would want to travel from one part of the cosmos to another via a wormhole if he knew that he could never leave?

Fifth, the gravitational tidal forces experienced by a traveler must be as small as possible. Nobody wants to be stretched out like a rubber band, but tidal forces have just that effect. (The gravitational tidal forces of the moon stretch the ocean a bit at high tide, for instance.)

Similarly, the acceleration and deceleration of the passengers must be kept to a minimum. Everyone knows the unpleasant effects of starting or stopping too quickly. It is true that on a roller coaster these effects can be exhilarating. But the thrill would certainly be outweighed by the discomfort if the motion were far more erratic and rough.

What would be a reasonable maximum level for the acceleration of a ship through a wormhole? The change in speed experienced by people or objects freely falling to earth from a certain height is called the earth's *gravitational acceleration constant*, often abbreviated as g, which is thirty-two feet per second per second. This means, for instance, that a woman diving from a diving board would increase in downward speed by thirty-two feet per second each second of her fall. Human beings can experience greater accelerations without much physiological damage, but there is a limit above which severe problems occur. Astronauts and high-altitude pilots endure several times the acceleration of gravity as they climb; they must be specially trained for such demanding circumstances. Morris and Thorne suggest that the earth's gravitational acceleration should be the limit for wormhole travel. Anything much more than that, they feel, would lead to severe discomfort and might be dangerous.

The sixth of Morris and Thorne's conditions refers to the time taken by passengers in crossing the gateway: the trip's duration according to the crew and to those left behind on earth. (Remember these need not be the same; for high-velocity travel, special relativity mandates that voyagers' times are dilated.) Here there is a rather strict requirement: the voyage must take less than one year for both the travelers and the people waiting behind. The reason for this stipulation is clear: space journeys of a few years or longer would be either dangerous, unhealthy, or tedious and might require costly procedures such as cryonic suspension. One also wouldn't want to travel through a wormhole and return home only to find out that one's loved ones had aged or died. Hence it is vitally important that the length of a cosmic gateway voyage be short according to both the spaceship's and the earth's clocks.

The seventh requirement is related to the creation of the wormhole and turns out to be the hardest to meet even if all of the others are met. It states that the matter and energy needed to form the wormhole must be physically reasonable; that is, they must possess the standard physical properties of known substances. Since the distribution of mass in an area determines its spatial and temporal geometry, to create a wormhole of a certain type one must stipulate the material arrangement in its region. A black hole needs a mass greater than the Chandrasekar limit (see Chapter 3) contained within a small enough surface; a traversable wormhole requires much more than that. We shall see in the next section that it might take an exotic new form of matter to create these special gateways; it is hoped that this construction could be accomplished with ordinary matter.

The eighth condition is that the gateway remain stable under the influence of the mass of the spaceship; that is, the gravitational influence of the vessel passing through the wormhole should not be strong enough to close the wormhole.

Finally, the ninth condition concerns the assembly of the wormhole. It must be possible to complete it in a reasonable amount of time, certainly much less than the age of the universe, and with a reasonably limited amount of matter, certainly much less than the material content of the universe. Further, it is hoped that some future civilization with advanced technology would be able to create an interstellar gateway by using the raw materials available at that time.

Basically, these nine conditions serve several different purposes. The first four and last two requirements concern the essential ingredients of wormhole creation: what makes a wormhole a wormhole. Conditions five and six pertain to the physiological needs of the human passengers on a spaceship traveling through such a gateway. Finally, condition seven relates to the reasonableness of the physical properties of the matter that makes up the wormhole. We shall see that this last requirement turns out to be the trickiest.

Exotic Matter

Essentially what Morris and Thorne designed was a "litmus test" for wormhole candidates. If a contender failed to meet any one of their nine criteria, then it should immediately be rejected, they reasoned. This had the effect of drastically reducing the number of theoretical celestial objects that could be classified as traversable wormholes.

On the basis of both these criteria and the equations of general relativity, they came up with a full description of the matter and energy required for a cosmic gateway to be fully realized. They calculated the properties of the matter needed at the throat of the hole and estimated the amount of its tension.

Essentially, tension is the breaking strength of an object. For example, the tension of a steel rod is the force needed to pull the rod apart. All objects have a point beyond which applied forces would destroy them and thus have fixed quantities of tension.

Unfortunately, Morris and Thorne found that the tension needed to prevent the wormhole's throat from caving in on itself would be enormous. The amount of tension was found to vary with the size of the throat. For example, for a throat that is four miles across, the quantity of force needed is roughly 10^{33} pounds per square inch—more than the pressure of a trillion pieces of cargo, weighing a trillion tons each, placed on a toy truck. In fact, this is about the same as the pressure in the center of a very massive neutron star.

If the throat were even smaller, the internal tension needed would be considerably greater. A throat that is the size of a football field (100 yards) across would require a pressure that is approximately 10^{38} pounds per square inch—10,000 times greater than in the four-mile-diameter case. These are all astronomically large quantities, far exceeding any pressures that have ever been observed in nature.

Finally, Morris and Thorne discovered a troubling fact about the matter needed for this gateway: the tension needed to prevent the wormhole from collapsing (expressed in pounds per square inch) must always be at least 10^{17} times greater than the density (mass per unit volume) of the substance used to build the wormhole. Note that 10^{17} is a little less than 1 million multiplied by 1 trillion. That is, for a wormhole constructed of matter of a density of one pound per cubic inch, the internal tension required must be at least 10^{17} pounds per square inch. (10^{17} is, in fact, the speed of light squared: the same factor found in Einstein's famous relation $E = mc^2$).

According to our current knowledge, however, no forms of matter existing in the universe today have breaking tensions so much greater than their density. Consider, for example, an ordinary steel rod, with a mass density of roughly 1 pound per cubic inch. Its breaking tension is about 100,000 pounds per square inch—much, much less than the million trillion pounds per square inch required for a wormhole.

When tensions and pressures of materials become so incredibly large, the laws of relativistic physics begin to show considerable strain as well. If the tension of a substance were to rise above the "magic" value of 10^{17} times its own density, scientists speculate that this material would begin to possess unusual properties. These include features, such as negative mass, that would defy all intuition about the nature of matter. For this reason, substances with tensions this great, called *exotic matter*, present us with the challenge of reformulating certain long-held physical notions.

What makes exotic matter, the substance of wormhole creation, so unusual? The key to this puzzling situation is related to the idea, in special relativity, that two observers can measure the same quantity (such as time, distance, and mass) and obtain two entirely different answers. For example, consider two football referees measuring the time taken for a football to be passed from one player to another. One of these referees, called Sammy, stands close to the field where the game is being played. The second referee, called Tammy, is situated in a high-velocity (near-light-speed) train, zooming past the action as it occurs. According to the time dilation principle of special relativity, Sammy and Tammy must record different values for the amount of time taken for the pass. Both of these results are valid; the difference is the result of the different speeds of the observers.

The same situation occurs in the case of mass. Suppose Sammy and Tammy both decide to measure the mass of the football and compare their results. Once again, just as in the case of time, special relativity dictates that they must obtain different answers for mass. Tammy, traveling close to the speed of light, thus perceives the football as being far more massive than does Sammy.

On the other hand, both Sammy and Tammy must agree that the football has *positive* mass. Under no known physical circumstances might either find the football's weight to be, let's say, negative ten (-10) pounds. In fact, no material on earth, measured under any set of conditions, has ever been found to have negative mass.

Exotic matter, that enigmatic substance that is absolutely *required* for wormhole construction, presents a puzzle: under certain circumstances, its mass would indeed be measured to be negative. Thus if the football players in our tale were tossing around a football made of exotic matter, Tammy, traveling in the high-speed train, might perceive it as a negative-mass object. Exotic matter appears to have negative mass when viewed by an observer moving at a sufficiently high speed. What ensures this unusual state of affairs? Einstein's theory of special relativity: the same set of principles that guarantees time dilation and length contraction.

Clearly, exotic matter confounds conventional scientific wisdom. Since no substance has ever been discovered to have negative mass, weighing a block of an unknown object and finding the scale to read −100 pounds would be most baffling; one would most likely conclude that the scale was broken. Yet that is exactly what would happen if a chunk of exotic matter were placed on a scale capable of recording negative mass and one were to fly by at just the right speed: the scale would have a negative reading.

A world with exotic matter of negative mass would be a strange world indeed. One is reminded of the children's story *Charlie and the Great Glass Elevator* by the late Roald Dahl. In one segment of this tale, Charlie and his grandparents discover a bottle of "youth pills," composed of a substance that gives the ingester renewed vitality. Charlie's grandparents become overly enthusiastic about the prospect of becoming younger and take far too many of these pills. As a result, they grow younger and younger until they are younger than when they were born. What happens to people with negative ages? Well, according to Dahl, they must sit in a giant waiting room until they are old enough to be reborn.

Does this mean that people who take too many diet pills eventually acquire negative mass and must then wait in a waiting room until they gain weight again? Are these unlucky people temporarily composed of exotic matter? Do people with negative ages also have negative mass? It is interesting to ponder these intriguing questions, ones to which Dahl certainly would have had satisfying answers.

Dahl's whimsical story helps to illustrate the speculative nature of exotic matter. Like negative ages, one would never have thought that masses below zero could exist. Yet this sort of unusual material, according to Morris and Thorne's theory, seems to be a necessity for the creation of traversable gateways.

Where would exotic matter be found? How would one go about looking for objects of negative mass? One possibility is to investigate unexplored regions of the cosmos or to use atom smashers to search for special negative-mass elementary particles of this sort. It is quite possible that our assumption that mass is always positive is simply a long-held prejudice and not an essential truth.

A far more promising approach to the search for exotic

matter lies in the realm of quantum mechanics. Physicists have long known that many events that could never occur in a Newtonian framework would be considered common-place from a quantum theory perspective. For example, if classical physics, according to Newton, were absolutely true on a subatomic scale, a hydrogen atom, consisting of a neg-atively charged electron orbiting a positively charged proton, would be unstable; the electron would spiral in toward the proton and collide. But we know that hydrogen atoms are indeed stable; otherwise, objects like the sun, composed mainly of hydrogen, couldn't exist. How can hydrogen's sta-bility be explained? Purely by quantum mechanics; certainly not by classical mechanics. So, we can clearly see that quan-tum theory can yield results that supersede Newtonian ex-pectations.

The first versions of quantum theory were developed by physicists such as Louis De Broglie and Niels Bohr in the early decades of the century. Although many successful pre-dictions resulted from their work, such as formulas repro-ducing the orbital levels of electrons in hydrogen atoms, it wasn't until the scientists Werner Heisenberg and Erwin Schrödinger appeared on the scene that quantum mechanics developed into the solid and respectable body of knowledge that is used today.

Heisenberg looked very carefully at the nature of mea-surement. It was commonly thought at that time that taking readings of quantities such as speed, position, energy, and momentum produced answers independent of the measurer: that merely observing a property of an object would have no effect on the object or its other characteristics.

In 1927, Heisenberg overturned these widely held notions. He proved convincingly that there exist pairs of quantities for microscopic objects about which exact knowledge of both at once is impossible; a measurement of one property renders a measurement of the second totally useless. An example of such a pair is position and momentum; for a given object, one can never obtain exact readings of both of these quan-tities at once.

To understand the *Heisenberg principle of uncertainty*, imagine a poor television set with two tuning knobs. One of these dials focuses the sound while the other tunes the picture. Suppose the uncertainty principle were to apply to the sound and vision of the set. If you were to tune in the sound for crispness and clarity, the picture would go fuzzy. If you tried to adjust the picture dial until the image was perfectly clear on the set, the sound would turn to static. Thus, you could never experience clear sound and vision at once; you would always have fluctuations of one mode or another.

Similarly, following Heisenberg's uncertainty principle, the properties of tiny objects are always fluctuating. In general, the position, energy, and momentum of microscopic particles vary over time in a way that is not entirely predictable. It is possible to predict the averages or most likely values of these quantities but never the *exact* values of all of them at once. And these random fluctuations make complete knowledge impossible.

Furthermore, these quantum fluctuations have measurable physical effects, properties that prove extremely useful in our search for the exotic matter needed to form wormholes. One of these is called the *Casimir effect*, named after the Dutch physicist Hendrik Casimir, who calculated it in 1948. Casimir imagined two metal plates placed a small distance apart. Normally, if an electric charge were placed on these plates, one would expect an electric force to be produced between them. And conversely, if there weren't a charge so placed, then there wouldn't be a force between them.

Casimir predicted that in the second case, even in the absence of net charge, there would nonetheless be a small, but measurable, force linking the plates. This force would emerge from the electromagnetic energy between the surfaces. But where would this energy come from if there is nothing between the plates?

Well, Casimir pointed out that there's no such thing as nothing! Heisenberg's uncertainty principle tells us that we cannot know the energy of any part of empty space exactly all the time. That means that this energy should fluctuate (just like television static) and that new particles should constantly be created and destroyed, emerging out of the vacuum and returning again to the vacuum after a brief interval in "reality." Thus "empty" space, rather than being seen as a tranquil sea of nothingness, should be viewed instead as a turbulent ocean full of transient particles and fluctuating electromagnetic fields.

Clearly this sea of *virtual particles*, as they are called, should have an effect on the surrounding plates. Indeed it does, and a small force can be measured between the plates exactly as Casimir predicted. Thus, somehow, nothing (a vacuum) has produced something (a force).

If one measures the average energy of the space between the plates one finds the interesting result that the fluctuating fields produce a value less than zero. This occurs because these fields are, on average, situated beneath the surface of the particle sea. Since, according to Einstein, energy and mass are freely convertible forms of the same quantity, the mass density associated with the vacuum is similarly less than zero. Hence the vacuum fluctuations in free space, predicted by the Casimir effect, can be construed as an example of negative mass.

Since negative mass or energy density is a hallmark of exotic matter, perhaps the Casimir effect could be of some use in creating the material needed for wormhole throats. Although this prospect is highly speculative, scientists who are looking for exotic matter view the Dutch physicist's remarkable result as a source of hope for the future of wormhole construction. Maybe the means to produce exotic matter are within our reach after all, and, to activate them, we simply need to use the power of quantum mechanics. For this reason, researchers today are busy trying to exploit the unusual properties of vacuum fluctuations such as the Casimir effect and tap into possible hidden sources of power. It is ironic that one of the storehouses to be tapped may be empty space itself.

Much to the chagrin of traversable wormhole enthusiasts, the possibility of the existence of exotic matter has long been dismissed by most physicists. This taboo against negative mass even has a name: it's called the *weak energy condition*. Yet, as the Casimir effect demonstrates, it is entirely unclear whether this prohibition against the matter necessary for wormhole construction has any basis in physical reality.

Some theorists even assert that the existence of the Casimir effect is sufficient to *disprove* the weak energy condition experimentally. So, even though the exotic matter requirement is troubling to wormhole designers such as Morris and Thorne, they certainly don't consider it an insurmountable obstacle to the full development of the theory.

Constructing Wormholes

How would an advanced civilization go about creating a navigable wormhole? What would be the talents needed to deal with the requirements of construction, the technical difficulties that might be encountered? Where would we find the exotic matter needed to create this gateway? And how could such a structure be kept stable?

Obviously there are a host of questions that need to be answered before transgalactic travel shifts from the drawing boards of speculation to the construction rigs of reality. Morris and Thorne's work is a promising start; let's see what would be needed to take the next step.

First of all, we would need either to discover or to manufacture exotic matter. One place where future civilizations might mine exotic matter is in the vicinity of a black hole. Stephen Hawking's finding that black holes decay (see Chapter 3) lends itself to some fanciful thinking about possible sites to search for this unusual sort of material. As we have seen, black holes evaporate because of particle-antiparticle pairs created near their event horizons. These matter-antimatter companions emerge from the vacuum, appearing quite literally out of nothing, as a result of the quantum fluctuations we can postulate via Heisenberg's uncertainty principle (the same fluctuations giving rise to the Casimir effect).

It has been proved that once these particle-antiparticle pairs are created, they have a limited lifetime. Normally they must reunite within a very short interval of time and return to the vacuum; otherwise the laws of physics are violated. Since these microscopic objects are literally created out of nothing, their total energy must add up to zero. This means that one member of a pair has a positive energy (and mass), while the other has a negative energy (and mass).

What happens, though, if the negative-energy particle falls into the event horizon (boundary) of the black hole? Certainly it can't escape and reunite with its companion. Therefore, the object of negative energy must continue its descent into the core of the black hole, while the positive-energy particle is left alone to travel away from the black hole. An observer standing near the black hole horizon would therefore detect a net release of positive particles away from the black hole and a net flow of negative energy and mass into the horizon.

This is how black holes decay; it is also a possible source of exotic matter. If somehow we could place a kind of "net" near the horizon to scoop up the negative energy just before it enters the black hole and could store this material, we would have a considerable reserve of exotic matter. This exotic matter would derive from the overall flow of negative mass into the black hole boundary, captured before it reaches the point of no return.

As an analogy to these somewhat abstract concepts, consider the unusual behavior of flying fish. Ordinarily, fish must stay beneath the surface of a body of water. However, certain species of fish can, for limited periods of time, jump out of a stream, providing they quickly return.

Now imagine a particle-antiparticle pair to be a set of two fish: one "positive" fish and one "negative" fish. These fish dwell in a vast stream (the vacuum) bounded on the left by a river bank (the edge of a black hole, that is, its event horizon). Every few minutes they jump out of the water as a pair; separate, with the negative fish going to the left and the positive one to the right; and remain in the air for a few seconds. Then they both return to the stream, entering the water a few feet from each other.

One time, however, the fish stray much too close to the riverbank. The negative fish rises out of the water and, as usual, separates from its companion, but this time it finds that it has headed too far left for a safe return to the water. Instead it must land on the riverbank, leaving its positive mate behind. Just before the negative fish hits the ground, though, it is caught by a fisherman sitting on the embankment. This fisherman is a collector of "exotic" negative fish and finds that he can catch a lot by waiting there for just these sorts of strays. Similarly, collectors of exotic matter can wait near the boundary of a black hole and collect the negative-mass particles that stray close to this edge before they are absorbed.

This example of a way of obtaining exotic matter is admittedly conjectural and obviously quite risky. It is hard to imagine future astronauts parking their ships close enough to black hole event horizons to dredge up the negative energy available there. Robot vessels would have to be used to navigate these perilous "waters" and perform this hazardous task. Still, even if we must relegate exotic-matter mining to the far future, it is encouraging to see that this possible avenue of exploration exists.

Clearly, the next step after collecting the exotic material would be to assemble the wormhole and ensure that its gateway is connected to the regions of the universe we wish to link. Unfortunately there is no way at present to build a structure of such magnitude and properly orient it. This doesn't mean that this task is impossible; ideally, future civilizations will be able to create objects as vast as asteroids, planets, and even wormholes.

So let's make a leap of faith and assume that advanced societies will indeed be equipped with wormhole building devices. The next hurdle with which we must reckon concerns the amount and distribution (structure) of the exotic matter to be used. Exotic matter, as we have seen, would be acquired as a resource at enormous cost and with considerable effort. We'd want to use as little as possible and spread it out in the most effective manner. Morris and Thorne suggest several schemes for reducing the quantity needed of this problematic substance; all of these involve concentrating the precious matter within a limited band around the center of the wormhole. In each of these schemes, ordinary matter composes most of the wormhole, while exotic matter forms a segment of the central part of the throat.

Another related issue involves shielding the passengers from interactions with the exotic material. Since this substance would be so strong that it would have the tension of a neutron star, it might have harmful effects on human beings. There are three possible ways to deal with this problem. One would be somehow to shield the travelers: they could journey through a protective tube, for instance. The second way would be to use special sorts of exotic matter that would have only limited effects on humans. The third means of dealing with the exotic substance would be to keep it as far away from the center of the wormhole's throat as possible. In this manner, the human passengers could, by remaining near the center, avoid getting too close to the extremely high tensions and densities of this sort of matter. Physicists hope that by using at least one of these methods the theoretical difficulties involving exotic matter interactions with passengers would prove to be manageable.

The final problem with building and maintaining wormholes concerns stability: how to make sure that they would remain permanent structures. This issue was successfully addressed by the Caltech group; they did find stable wormhole solutions to their equations. Unlike white holes, which rapidly decay on contact with normal matter, wormholes need not be affected by other materials. Nevertheless, permanent monitoring teams would be needed to make absolutely sure that any wormholes created didn't collapse. Small adjustments could be made as needed to ensure that the cosmic gateways would remain open for as long as they were needed.

The technical problems associated with wormhole creation and maintenance are formidable. We still don't know how to find exotic matter, to assemble it into a gateway, and to protect voyagers from its strong effects. But it is equally true that Morris and Thorne's article has stimulated a lot of interest—and work—in systematically addressing these problems so as to find comprehensive and workable solutions.

*T*he Space Mirror

In the spring of 1989 a high-priority report entitled "Traversable Wormholes: Some Simple Examples" appeared in a prestigious physics journal. If that journal were a tabloid, its headline might have read "NEW IMPROVED WORMHOLES DISCOVERED: INTERSTELLAR TRAVEL CLOSER TO REALIZATION." Despite the more restrained presentation, the appearance of the notice in the feature section of the journal provides some indication of the importance of the topic and the ingenuity of its author, Matt Visser, of Washington University, St. Louis. (Some of his work was done at Los Alamos as well.)

Visser is part of a new group of physicists specializing in astrophysics and cosmology, who are interested in the mechanics of wormhole construction. Inspired by the work of Thorne, Morris, and Yurtsever, these scientists wish to push the boundaries of our knowledge about interstellar gateways to their absolute limit. They have set out to modify the Morris and Thorne Caltech models to eliminate or reduce some of their outstanding problems, particularly with regard to the use of exotic matter in wormhole construction. The less one must deal with this troubling substance, the better, as far as these physicists are concerned. Therefore they wish to maximize the use of normal matter (with positive mass) and minimize the use of exotic matter (with negative mass).

Visser looked at Morris and Thorne's assumptions one by one and tried to figure out which could be altered without destroying the merits of their wormhole model. Scanning the list of conditions for wormhole construction that Morris and Thorne found essential, he concluded that the only one that seemed superfluous was the requirement for spherical symmetry. This condition had originally been placed in the list simply to make calculations easier for the scientists.

Visser, in fact, has developed a model without spherical symmetry at all. His wormhole solution appears nothing like a flattened hourglass; it looks more like a rectangular version of a spool used for storing thread. To understand Visser's model, picture an ordinary household spool, the cylindrical sort with a circular hole in the middle. Naturally, its top and bottom extend outward from the spool to prevent thread from unraveling. Now imagine that the spool is altered to become *rectangular*, with a rectangular hole in the middle.

This rectangular spool design provides a three-dimensional representation of Visser's four-dimensional model. In this portrait, the top of the spool corresponds to one part of the cosmos and the bottom to another part. These surfaces are flat, reflecting the fact that ordinarily, in the absence of matter, space is said to be flat (this is a consequence of Einstein's general theory of relativity). The only curved part of the spool is its rectangular center, which corresponds in Visser's model to the throat of the wormhole. The edges of this hole are lined with exotic matter, the substance needed for wormhole creation.

Now let's imagine how Visser's model could serve as a wormhole. The following is naturally a four-dimensional depiction: besides the three dimensions of normal space that one ordinarily imagines, there is a fourth, along the length of the wormhole throat. Spaceships would enter the wormhole traveling along the flattened top of the spool, a region which corresponds to one part of the universe. Next, they would enter the throat of the wormhole through the center of the spool's rectangular hole. Finally, after traveling through the throat, they would emerge on the bottom of the spool, which corresponds to another part of the universe. Thus the rectangular hole serves as a gateway between two regions of space.

Essentially, Visser's theory has a strong advantage over Morris and Thorne's. In Morris and Thorne's approach, the wormhole's throat resembles the narrow connection between the two basins of an hourglass. Naturally, if one were to pass through such a throat, from one basin to another, one would find that the throat's walls would become closer and closer to each other as the journey continued. Eventually, one would need to traverse the thinnest part of the throat, halfway between the upper and lower basins. Because the walls of this segment must be composed of exotic matter, and they are so close together at this point, one would need to travel perilously close to this exotic material.

On the other hand, in Visser's model the walls of the

wormhole throat are straight-edged and can be made as distant from each other as one likes. Therefore, travelers passing through the wormhole could avoid the boundaries of the throat altogether and remain at all times in the middle of the hole. So, the inner edges of the throat could be lined with exotic matter, but nobody need ever be exposed to it.

Visser's model would, most likely, be much more stable than Morris and Thorne's. Because of the lack of contact with the exotic material in Visser's model, one wouldn't have to worry as much as with Morris and Thorne's about damage to the wormhole's structure. For the same reason, safety would be much greater with Visser's; gravitational tidal forces on passengers would hardly amount to anything at all.

Creating this sort of wormhole, as in assembling the Caltech model, is still beyond the realm of current technology. In that sense, Visser's model is just as hypothetical as Morris and Thorne's. Yet it is also true that the decreased reliance on exotic matter places his model a step above their earlier blueprint.

How would Visser's wormhole appear to travelers as they entered it? Certainly not as a spool; the spool representation is a four-dimensional picture of the wormhole—it is the image that one would see if one were to step outside normal space altogether and look at the wormhole in its entirety. What space explorers would actually see would be a three-dimensional cross section, a slice of the total picture. This slice would consist of a three-dimensional rectangular hole cut out of space (corresponding in the spool model to the top of the central rectangular aperture). How would a rectangular hole in space appear? Basically, it would look like a dark, rectangular box, floating in the cosmos: a sort of black prism. A starship would approach the prism and enter it via the center. After a brief interval, it would reemerge through a similar dark prism in another part of space. These prisms would be connected via the fourth dimension along the shaft of the spool, namely the wormhole's throat.

The parallels between this theoretical depiction of a rectangular wormhole and Clarke's fictional portrait of the Star Gate in *2001: A Space Odyssey* are truly extraordinary. Consider the following passage, describing David Bowman's entry into the Star Gate:

> [Bowman] literally could not describe what he was seeing. He had been hanging above a large, flat rectangle, eight hundred feet long and two hundred wide, made of something that looked as solid as rock. But now it seemed to be receding from him; it was exactly like one of those optical illusions, when a three-dimensional object can, by an effort of will, appear to turn inside out—its near and far sides suddenly interchanging. . . . What had seemed to be its roof had dropped away to infinite depths; for one dizzy moment, he seemed to be looking down into a vertical shaft.
>
> David Bowman had time for just one broken sentence. . . . "The thing's hollow—it goes on forever—and—oh my God!—it's full of stars!"

Compare this description to Visser's far more sober ideas (in *Physical Review*, 1989):

> It is now easy to see how to build a wormhole such that a traveler encounters no exotic matter. Simply choose (the throat) to have one flat face. . . . A traveler encountering such a flat face will feel no tidal forces and see no matter, exotic or otherwise. Such a traveler will simply be shunted into the other universe. . . .
>
> [For this] any rectangular prism would do just as well.

One can imagine Bowman meeting with the alien designers of the Star Gate, asking them the secret of interstellar travel, and hearing them respond, "Any rectangular monolith would do just as well." Indeed fact here seems to mimic fiction.

Visser, in his article, points out another interesting feature of his wormhole—its exterior acts as a giant mirror. A beam of light aimed at its rectangular surface would bounce off just as if it had hit a reflecting glass. As in the case of a plane mirror, the beam's angle of incidence would be exactly the same as its angle of reflection. However, the incident beam would enter from one region of the universe, while the re-flected beam, having passed through the throat of the worm-hole, would be sent into another sector of the cosmos. And this, of course, marks an important distinction between a "space mirror" and an ordinary looking glass.

It's somewhat amusing to note the parallels between hav-ing one's reflection end up in another part of the universe via wormhole transmission and the fairy tale notion of step-ping through a mirror into a "looking glass kingdom." Per-haps Lewis Carroll was more prophetic than he might have thought!

*S*chwarzschild Surgery

After developing his rectangular "mirror" wormhole, Vis-ser decided to construct another gateway model. Visser's second proposal, although less elegantly simple than his first, stands as a significant alternative design for an inter-stellar portal. Like the rectangular construct described in the previous section, it requires minimal use of exotic matter, is stable, and is fully traversable by human voyagers.

The idea for Visser model number two is similar to a common surgical procedure applied to organically grown apples. As anyone who has tasted natural, nonsprayed ap-ples knows, these fruits generally have a number of blem-ishes, including external bruises, wormholes (not the cosmic type but the sort caused by worms), and rotten parts. Usually, one simply trims off the blemished portions with a sharp knife, leaving the part of the apple that is suitable for con-sumption.

In his "Schwarzschild surgery" model, Visser observes that one of the principal problems with passage through an interstellar gateway between two Schwarzschild black holes is the existence of event horizons blocking escape from either terminus. This is the rotten part of the apple, so to speak. Visser suggests solving this problem by applying a cosmo-logical scalpel to each of the two Schwarzschild solutions, cutting out the unsavory parts: the event horizon and the central region. He then recommends grafting the external parts of each solution together via a central shaft (the shaft connects these regions along the fourth dimension). Thus the inside of each remaining "husk" would constitute a boundary of the wormhole throat.

This is somewhat hard to picture, since it is a higher-dimensional image; one might imagine two coreless apples in which the inner walls of the fruit are connected along the fourth dimension. The basic idea is that one would enter through the skin of one of the Schwarzschild models, travel through the throat, and then emerge through the skin of the second one. Therefore, without event horizons blocking passage from the second Schwarzschild model, escape from this region would pose no problem and travel to another part of the cosmos would be theoretically possible.

The exotic matter of this wormhole would be concentrated in a thin layer on the boundary walls of the throat. As in the case of the space mirror, this concentration of the negative-mass material would help to eliminate problems caused by exposure to the substance, such as strong gravitational tides. Thus, in terms of minimizing the use of troublesome exotic matter, both of Visser's models are superior to Morris and Thorne's.

Both of Visser's models—the monolithlike rectangular slab scheme known as the space mirror approach and the surgically altered Schwarzschild solution known as the coreless apple model—seem theoretically viable, assuming the exotic matter problem can be solved. However, wormhole construction crews of the future might well prefer one approach to the other for purely pragmatic reasons.

The World Turned Inside-Out

The wormhole models developed by Morris, Thorne, and Visser provide a wealth of information about what journeys through such interstellar gateways might be like. They yield much useful knowledge about how passengers might *feel* during such voyages: how much acceleration, gravitational tidal force, and pressure they might experience, for instance. It is interesting to probe these models for further information about wormhole journeys; what might travelers actually *see* as they passed through these space tunnels? That is, if passengers on a wormhole-bound vessel turned around and looked out the window, how would the distant stars appear?

Since Morris, Thorne, and Visser's models are so recent, no direct calculations have been made, as yet, of the appearance of distant objects to an observer in a wormhole space-time. However, recently an article by the Australian mathematician William Metzenthen has shown how objects would appear to travelers journeying into different sorts of black holes. Metzenthen's conclusions could perhaps be extended to traversable wormholes as well, so long as one is careful to deal adequately with the differences between black holes and wormholes. They paint an intriguing picture of what astronauts would see if they were traveling through an interstellar gateway.

Imagine a starship soaring toward a wormhole in a free fall motion. Suppose the walls of the ship are made of thick, transparent glass and it is possible to look outside at space, observing the sky during the flight. Any changes in the appearance of the stars and the constellations can be readily noticed.

At first, while the spaceship is thousands of miles from the wormhole entrance, space appears normal. The stars seem to shine with their normal intensities, exhibiting their usual spectrums of colors, and are spaced at typical intervals. Aside from the fact that the constellations and stellar patterns are different, the sky looks exactly as it would on earth.

However, as the ship approaches the throat of the gateway, something odd begins to happen. Gaps start to appear in the sky as the stars begin to assemble into bands of light. Then these bands begin to replicate themselves, spreading out across the heavens in multiple images. Any given constellation, instead of appearing just once, is seen as an ever-growing set of facsimiles. The sky increasingly seems to be reflected in a set of peculiar funhouse mirrors.

This multiplication of images is related to a recently discovered phenomenon called *gravitational lensing*. If a man looks at his reflection in a curved mirror, the image appears distorted. Depending on the mirror's curvature, the reflection is either shorter and squatter than the man, taller and lankier than he, or altered in some other way. This distortion results from the scattering of the light rays reflected from the man on hitting the mirror's surface and being further reflected. Similarly, if an object is viewed through a curved lens, the object appears changed, also because of the bending of its light rays. Here the slowing of light as it passes through the material of the lens causes the alteration of its path. Finally, in *gravitational lensing*, gravity causes the bending of light, leading to a distortion of images. The curvature of space itself, because of the presence of mass, causes light in a certain region either to spread out or to focus itself into a bundle. The resulting visual effect is either a stretching or a multiplication of images.

Let's consider how this works in the vicinity of a wormhole. The extraordinary mass of the wormhole causes the light rays in the area to concentrate and focus themselves into clusters. Suppose these rays are produced by stars. Normally, the light produced by stars appears in straight lines, and the eye traces these lines back to single points. Thus, the stars appear to be single entities. Near a wormhole, however, light rays *converge*. The human eye traces these lines backward to obtain a picture of the sky. Convergent lines traced backward are divergent: they seem to split apart as they retreat from the eye. Thus, the light from the stars appears to be split into discrete bundles throughout the heavens. The picture obtained is then one of multiple stellar images scattered about the sky—hence the funhouse mirror effect.

As the spaceship nears the wormhole, other distortions of the sky's appearance take place as well. A shifting occurs in the apparent color spectrum and intensity of stellar light as it reaches the ship. First of all, the wavelengths of the light waves appear shorter and shorter: reds become oranges, oranges switch to yellows and greens, while greens shift more and more into the blue and violet end of the spectrum. This wavelength-decreasing process, the opposite of a redshift, is known as a *Doppler blueshift*, and is a phenomenon normally associated with either high velocity of the light source or strong gravitational fields. In the case of wormholes, tremendous gravitational forces cause the incoming light to increase in momentum, resulting in a frequency increase and a wavelength decrease.

Corresponding to the blueshift is an increase in the energy and intensity of the perceived light. This growth in intensity manifests itself in the apparent brightness of the light. Hence, as the craft nears the wormhole, the light from the sky appears more and more fractured, increasingly blueshifted and continuing to grow in strength.

Clearly, wormhole passage would provide voyagers a most intriguing light show. The specific nature of the sky's distortion would depend to some extent on the type of wormhole involved. Morris and Thorne's model, for instance, would produce one sort of imaging and Visser's would produce other patterns. But regardless of the wormhole design, the sights seen by travelers en route as they gazed out of their spaceships' windows would be most unusual indeed —a veritable collage of color.

A Tangled Web

Some of the most imaginative ideas about traversable wormhole creation stem from an extremely theoretical branch of physics: the science of quantum gravity. *Quantum gravity*, the hypothetical link between the theory of general relativity and the laws of quantum mechanics, has emerged in recent years as one of the prominent fields of inquiry in physics. Theorists in this area seek to enlarge our understanding of the forces of gravitation, especially on a small-scale level. Within this domain, the subject of wormholes has emerged as a by-product of discussions about the deepest levels of space-time in attempts at formulating a quantum theory of gravity. It is hoped that microscopic wormholes uncovered by such a theory could be enlarged somehow and used for interstellar space travel. Instead of having to build these gateways from scratch, they could simply be salvaged from space using quantum gravity theory as a guide.

Within the microscopic world of particles, gravity is the

least understood of all of the natural forces (strong, weak, electromagnetic, and gravitational). Over the last few decades, comprehensive theories of the electromagnetic and weak forces have been developed and tentative notions about the nature of the strong force have been discussed. Together with gravity, these form the four fundamental particle interactions, and all natural events comprise combinations of one or more of these couplings.

In the macroscopic world, with which we are most familiar, quantum mechanics plays only an indirect role. Operating in the realm of the tiny, however, one finds that quantum effects are pivotal. The atomic and subatomic scales are ruled by the mysterious probabilistic domain of the physics of Schrödinger and Heisenberg. Randomness and *information smearing* (the spreading out of physical quantities over a range of values) are the hallmarks of this chaotic and only marginally understood region.

Quantum dynamics is governed by the Heisenberg uncertainty principle, which rules out immediate and complete knowledge of a particle state (the set of all of a particle's properties). As discussed, properties such as momentum and position cannot be known at the same time for a given particle. As in the analogy of the television set, adjusting one parameter (momentum, for instance) to make it sharp and focused makes the other one (position) fuzzy. What, then, *can* we say about a microscopic object?

Instead of talking about *observable* quantities such as energy and location, the laws developed by Schrödinger and Heisenberg mandate that we consider as fundamental an abstract object called the *wave function*. All of the information about a particle for any given instant is buried within the wave function, yet, in accordance with the Heisenberg uncertainty principle, not all of this information can be retrieved at once. When scientists take measurements of particular quantities, they glean specific answers only if the questions are "compatible" according to this principle. Measuring position and momentum at once is an example of incompatibility; both pieces of information cannot be calculated from the wave function simultaneously.

Suppose one measures one of two incompatible properties; what can one say about the value of the other property? According to quantum theory, at that instant the value of the second quantity is "fuzzy"; that is, it is spread out over a range of possibilities. In fact, the second property's value *stays* fuzzy until it is measured. This illustrates a fundamental difference between quantum and classical mechanics: physical properties of a particle form a spectrum of possible values until they are measured by an observer. Then, after the measurements are taken, these properties take on one of a range of values. The wave function, at that time, is said to *collapse* into a definite state representing this one particular value of the quantity being measured.

For example, consider a measurement of the location of an electron. Before the electron's position is scientifically observed, it cannot be said to have a definite value. After a position measurement takes place, the electron's location becomes precise as its wave function collapses into that for one particular site. Consider another example: electron spin. An electron can spin either counterclockwise (called *spin up*) or clockwise (called *spin down*). Until the electron's spin is measured, its spin is said to be in a mixed state, neither up nor down.

This strange property of microscopic phenomena gives rise to some fascinating philosophical issues, the most famous of which is Schrödinger's cat paradox. Here it is imagined that a cat is placed in a black box and somehow attached, by electric wire or otherwise, to a spin detector. An electron is sent into the spin detector and the spin direction is measured. If the electron is spinning so that its axis is pointing up, the cat is killed. If the spin is down, then the cat is spared. The entire setup is covered with a black cloth so that no measurements can be read nor can the cat be seen.

The paradox of this rather morbid experiment is as follows: Since no measurements of the electron's spin have been taken (or at least a scientist hasn't seen the results), the electron must remain in a mixed state, part up and part down. That means that the cat is also in a mixed state, half alive, half dead. Only when the black cloth is removed and the box opened can one definitely say that the cat has lived or died. The cat remains in a sort of quantum limbo, waiting for its wave function to collapse into a definite state. Clearly this cannot happen in real life; the cat is either functional or deceased. However, contrary to our expectations, quantum mechanics dictates otherwise.

Because of such perplexing theoretical ambiguities, the standard interpretation of quantum measurement theory (that wave functions are "fuzzy" until they collapse when a scientist takes a measurement) remains controversial more than sixty years after its formulation. However, quantum mechanics produces excellent agreement with experimental evidence about many matters and is not to be taken lightly. Until particle phenomena can be explained by an even more accurate approach, we must assume that quantum theory, however puzzling, is absolutely correct.

Quantum gravity is the last frontier of quantum mechanics and, some believe, the last frontier of particle physics. This attempt to apply the Heisenberg uncertainty principle to the domain of space-time and gravity began in full force over thirty years ago after it became clear that electromagnetism can be fully quantized (expressed in terms of quantum theory). Therefore, physicists have surmised, it is likely that gravity can be quantized as well. Unfortunately, a successful theory of this sort has not yet been developed.

What might a quantum theory of gravitation be like? First of all, it would apply on an extremely tiny scale, at lengths of 10^{-33} inch or less. At this level, the geometry of space and time would look very strange. Because of the uncertainty principle, space-time would appear turbulent and inexact. Since quantum mechanics would guarantee lack of precision in measurements, geometries would fluctuate among several possible configurations—flat, toroidal, and saddle-shaped, for example.

Theorists such as John Wheeler have referred to the nature of the universe on such a small scale as consisting of *space-time foam*, a multiply connected, turbulent sea of geometry. Connections among separate parts of the cosmos would take place on this level via microscopic wormholes. These miniature wormholes wouldn't need to be constructed like the large-scale ones; rather they would occur naturally in the space-time foam because of the complex interconnections of geometric entities. The uncertainty principle would guarantee a lack of knowledge about these possible wormhole links; therefore, they might all take place.

In other words, just as in the case of position, momentum, or spin, there would be a "fuzziness" built in to the nature of space-time connections themselves. In standard general relativity, any two points define a "shortest" connection between them: the path that light takes between one point and the other. The set of all these connections, the metric, is well defined and can, in theory, be determined with great precision. On the other hand, in quantum gravity no such exact determination could be made. The metric of the universe would fluctuate among several configurations and wormhole shortcuts would constantly be created and destroyed, leading to an oscillating range of possible paths between any two points.

Thus, we can see that there is a sharp contrast between Einstein's model of space-time and the newer "spongy" model of quantum gravity. In the former, space-time would be well defined down to all levels, including the microscopic. In the latter, the quantum approach, the universe at its deepest level would be a maze of wormhole links, enabling all possible paths through the cosmos to be realized with varying probability. Uncertainty would be king in this nether realm.

Morris and Thorne have speculated that this "space-time foam," with its tangled web of connections, would be an ideal place for a powerful future civilization to search for, find, and then enlarge wormholes. Such a society could pull a microscopic wormhole out of this cauldron of geometric possibilities, increase its size to macroscopic dimensions, adjust its properties to fit the safety regulations they've established, and then move it to an appropriate part of the cosmos, where it could be used in its planned role as an interstellar gateway.

These remarks are, of course, highly speculative. Today it is impossible for us to imagine ways of accessing the theorized space-time foam. Indeed, we aren't even fully sure of its existence, since the whole subject of quantum gravity still

is in its infancy. But it is truly amazing to ponder the possibility that powerfully advanced cultures, using tools and
methods at least conceivable today, might someday learn
how to make use of the deepest levels of the cosmos, molding
and shaping its underlying essence to fit their needs.

CHAPTER 6

THE BLACK HOLE MINING
EXPEDITION OF 2500

*T*apping Hidden Energies

The major part of the universe consists of empty space, a vast dark nothingness that no one can fully fathom. Yet theories of quantum gravity suggest that hidden within the deepest recesses of this monumental vacuum lies a tremendous storehouse of energy, a colossal power source that could serve all human needs for millions of years. All that is needed is a mechanism able to tap into this source of power. If there is indeed a way to achieve the ultimate dream of getting something from nothing, wormholes, with their ability to breach seemingly impenetrable barriers, will play a tremendous role in this quest.

At present, the human population is growing at a tremendous rate, doubling every few decades. Moreover, the world's rate of fuel consumption per person is expanding as well; new technologies are requiring a larger and larger share of the world's resources. This all translates into a frightening, yet unmistakable, fact: the world's energy use is increasing exponentially with no end in sight.

Earth's energy resources, on the other hand, are finite. As the crises in the Middle East have shown us, oil reserves are limited and increasingly expensive to extract. Within the next century, at current rates of consumption, these supplies will be effectively exhausted. Coal and other fossil fuel reserves will hardly fill the gap; they'll be used up a few centuries later. Nuclear fission is problematic; it can be the source of life-threatening radioactive waste, as in disasters such as Three Mile Island and Chernobyl. Besides, like fossil fuels, uranium and other nuclear fuels are exhaustible commodities.

Sadly, *nuclear fusion*, a method for extracting energy from abundant hydrogen, remains an unproven source of usable power. In current fusion experiments a tremendous amount of heat energy (that is, a high temperature) is needed to initiate the fusion process and relatively little energy is generated. Recently, some have postulated that *cold fusion*, fusion performed at room temperatures without much energy input, might be possible under the right circumstances. However, after an exhaustive search for proof that cold fusion can take place, no evidence has been found to support this idea.

It's true that some of earth's energy sources are indeed "renewable." Solar energy, wind power, hydroelectric power, and geothermal energy are all naturally abundant and seemingly perpetual reserves of power. If these resources could completely fulfill our energy needs, then we wouldn't have to worry about running out of supplies, at least for a few billion years.

But what if a future civilization were to need more energy than the sun or the earth could provide? And what if, in a horrific catastrophe, the sun itself finally were to expire, its nuclear fuel exhausted? Citizens of a future society would be faced with a dire choice: find new extraterrestrial sources of power or perish in an icy demise. Traversable wormholes would, perhaps, provide the escape routes needed for these frightened individuals to relocate to more energetic regions of the cosmos or maybe the funnels required for them to carry energy and matter from other parts of the universe back to the earth.

In any case, to summon up the extra reserves of power needed to conquer space or simply to ensure the survival of humanity, researchers of the future would most likely turn to the hidden energy "frozen" into the fabric of the universe. Even in the absence of a comprehensive theory of quantum gravity, modern physicists have a good idea where these power cauldrons might lie. Astronomical measurements produced by present-day observatories are providing more and more precise maps of the energy and matter content of the universe, guides that will help future prospectors to zero in on the most lucrative fuel sources.

Scientists now know that the visible stars and luminous gases that speckle the nighttime sky make up but a small part of the matter and energy of the cosmos. Until very recently it was thought that most of the universe consisted of galaxies, themselves containing stars, planets, gases, and floating interstellar debris—in short, objects detectable by ordinary telescopes, albeit powerful ones. Today, however, the evidence overwhelmingly suggests that over 90 percent of the material and energy content of the universe is *invisible*. That is, most of the mass of the cosmos is in the form of *dark matter*.

What is the evidence for dark matter? A vital clue concerns the velocities of stars moving around the centers of galaxies. As in the case of the solar system, all galaxies rotate, and the stars constituting these galaxies orbit the galactic centers at detectable speeds. There are some similarities between the paths of stars around the middle of a galaxy and the orbits of the planets around the sun, but there is one crucial difference: whereas the outer planets are slower in their orbits than the inner ones, the outermost stars in a galaxy have the same orbital speeds as the innermost stars. In other words, although Pluto, the outermost planet of the solar system, travels much, much more slowly than Mercury, the innermost planet, the stars in the outer reaches of the Milky Way galaxy are just as rapid in their orbits as stars near the center. This distinction between the behavior of the constituents of solar systems and galaxies points to a strong difference in their mass distributions. In the case of the solar system, the drop-off in speeds away from the center means that most of the mass is centrally concentrated. This is clearly the case; the sun, in its central location, is far more massive than the distant planets. In the case of galaxies, the *absence* of a drop-off in speeds means that most of the mass of a galaxy lies away from its central nucleus. This is in complete contradiction to the standard picture of galaxies: massive central cores surrounded by thin discs. How can one then account for the unobserved missing mass? The answer seems to be that most of the mass of galaxies is present in the form of dark matter.

Other evidence points to the prevalence of dark matter in the universe. According to the big bang theory, the galaxies are receding from each other at a rate that can be determined by astronomical measurement, and, depending on its total mass, the universe will either expand forever or collapse back to a point. Current measurements of the *visible* mass of the universe suggest that there is much less matter than required for recollapse. However, detailed estimations of the recession of galaxies tend to support the opposite view: that recollapse is likely (after an extremely long period of time). The discrepancy can only be resolved by assuming that *dark* matter dominates the cosmos and slows the galactic recession.

Finally, strong evidence for dark matter occurs in the form of calculations of the gravitational force present in galactic clusters. Galactic clusters are stable collections of galaxies; since they are stable, their total gravitational attraction can be determined. Once the total gravitational pull of an arrangement of objects is known, its mass can be roughly estimated. Calculating this value for all known galactic clusters yields the surprising result that the masses of the clusters are far greater than the masses of the galaxies they contain. Hence, a considerable proportion of the matter in galactic clusters is located *between* the galaxies. Most of this intergalactic material is unobservable; thus it can be considered dark matter.

Scientists are baffled by the concept that so much of the matter in the universe is invisible and have so far failed to explain this puzzle completely. There are a number of guesses as to the nature of dark matter; most of these theories involve postulated new forms of particles that inhabit the void. Another distinct possibility is that black holes of various sizes make up a large percentage of the missing mass.

It's clear that these black holes, formed during different eras of the universe, would contain vast supplies of energy as well. Truly, then, the case for dark matter suggests that the visible stars are far from the only power sources in the universe. Black holes and other hidden objects may in fact be the last frontier of energy exploration.

Black Holes as Energy Sources

When the great railroad networks were built across the United States during the nineteenth century their purpose was twofold: first, to transport passengers; second, to carry freight. In particular the "iron horses" carried (and continue to carry) coal from mines in Pennsylvania, West Virginia, and other ore-rich states to towns and industries across the country.

Similarly, when wormhole transport networks or other interstellar gateways are established across the galaxy, they will also serve dual purposes: to carry people and to convey goods. Both raw materials and finished goods will be shuttled through these cosmic shortcuts from remote parts of the universe to earth, and out again, perhaps to other distant regions. Like the coal-carrying freight lines, the wormhole paths will be chosen, if possible, so that they adjoin the fuel-rich areas and provide for easy carriage back to the civilized zones.

The leading power sources of the nineteenth and twentieth centuries consisted of fossilized plants, their energies frozen into coal and oil strata. It is ironic that the power sources of the future may very well be found in other sorts of fossils, black holes. Coal and oil consist of solar energy stored as chemical energy; black holes, on the other hand, comprise stellar energy converted into gravitational and rotational energy in the form of fossilized stars. With wormhole transport systems taking them within reach of earth by spacecraft, black holes will very likely inspire the same sorts of mining expeditions that "black gold" spurred in the past centuries.

How could energy be obtained from a black hole? One possible method, to park a spaceship near the periphery and dredge for exotic matter, was discussed in the previous chapter. Another possible approach involves the draining of a black hole's rotational energy. As we have seen, one significant class of black holes, called Kerr models, exhibit rotation. Therefore, although some of their energy is stored in gravitational form, a large portion of their power is locked up in their spinning motion. By slowing this rotation, a tremendous amount of energy can be released.

This is similar to the case of a baseball player at bat who transfers the rotational energy of his body to the linear motion of the baseball. When a batter prepares for an incoming pitch, he twists and turns his body by at least ninety degrees. Then, as the pitcher throws the baseball, the batter creates spin energy by rotating toward the approaching projectile. Finally, when the batter hits the ball, he stops his turning and converts his rotational energy into a translational force, which, in turn, hurls the ball outward, away from the bat. The batter's spin energy has decreased by the precise amount that the baseball's energy of propulsion has increased.

A somewhat similar method for extracting spin energy from a black hole was developed by Roger Penrose. In Penrose's scheme, a spaceship approaches a rotating black hole with its engines shut down. It nears the event horizon but does not enter; instead, it moves around the black hole in the direction of its rotation. Once it is in a sufficiently close orbit, the pilot turns on the rocket engines. As the engines blast, the exhaust fuel is propelled into the event horizon. Since every action has an equal and opposite reaction, the release of the exhaust pushes the ship into a different orbit. In this new trajectory, the rotational energy of the black hole pushes the ship outward at a tremendous speed in a sort of "slingshot effect." Or, to extend the baseball analogy, one might say that the gravitational field of the black hole "bats" the ship out into space after an initial "windup" (the black hole's spin). The rotational energy of the black hole is thereby transferred, in part, to the spaceship and the black hole slows by a small amount. The ship in turn carries off a significant quantity of usable energy. Eventually, all of the spin energy of the black hole could be carried off in this manner.

In a straightforward extension of Penrose's scheme, we

can imagine the use of a rotating black hole, coupled with a well-placed wormhole, as an extremely powerful vehicular accelerator. Suppose that a traversable wormhole is aligned such that one of its mouths is close to the event horizon of a spinning black hole and its second mouth is located a few hundred miles from the first. Now picture a spacecraft parking itself near the horizon (and the first mouth of the wormhole) with its engines off. At the right moment, engines are turned on, the rocket exhaust is released, and the black hole "bats" the rocket into space with considerable velocity. This process is timed such that the ship is "batted" several hundred miles away from the black hole and then into the *second* mouth of the wormhole. Naturally, the craft re-emerges through the *first* mouth of the wormhole, where it turns on its exhaust again and is "batted" away again. Each time the ship completes this circular process, it picks up considerable speed from the black hole. Thus, the spaceship accelerates more and more, transferring an increasing amount of energy away from the gravitational vortex. Finally, when the ship has reached the desired speed, computers on board steer the ship away from the wormhole toward its final destination.

Every time the black hole is used to accelerate a star ship, the collapsed star slows in its spinning by a small amount. Thus, this process can be used only until the rotational energy of the black hole is drained. Assuming, however, that the black hole has started out spinning very rapidly, its lifetime as an acceleration device would be quite long—millions of years.

So far, we have shown how to milk a black hole of its rotational energy, to slow it down and transfer its spinning force to external objects. However, we haven't indicated how best to use the collapsed star's considerable internal gravitational fuel. This gravitational reserve arises from the steady contraction of the dead star down to its final black hole form. It is the same sort of energy possessed by the liquid in a water tower high above a building. Once the valves are released and the spigots turned on, the gushing power of the water flowing down through the edifice indicates the strength of its gravitational potential. This potential stems from the height difference between the tower and the base of the building and is exploited by the falling of the water. Similarly, a black hole contains a tremendous gravitational reserve just waiting to be released by the appropriate spigot. One simply has to devise a method by which the strong gravitational potential drop engendered by the black hole can be used.

One possible approach to exploiting the gravitational energy reserve of black holes is to merge two of them. When two such objects join, they create a resulting celestial body of greater mass and surface area than either of the original black holes. However, the mass of the final black hole is somewhat smaller than that of the sum of the initial components. The extra mass, in correspondence with Einstein's prescription that mass and energy are two sides of the same coin, becomes converted into usable power.

The amount of energy that can be extracted by this process depends on the nature of the collapsed stars. According to the British mathematician John Taylor, if neither of the black holes is charged or spinning, then almost one-third of the mass of the two objects can be converted into energy. This yields an efficiency of roughly 30 percent. If, on the other hand, four electrically neutral, nonrotating black holes are combined, this efficiency can be increased to 50 percent. In fact, the greater the number of initial separate objects, the more mass that can be converted into fuel. Since there is supposedly an abundance of these bodies in the center of the galaxy, a virtually unlimited amount of energy can be produced by black hole fusion.

For charged or rotating black holes, the statistics are even better. Two oppositely charged or spinning black holes can convert about half of their total mass into energy if they coalesce to form a neutral single body. Once again, as in the case of neutral black holes combining, by increasing the initial number of objects, the efficiency rises significantly. For instance, if eight spinning black holes are combined, roughly 90 percent of their mass can be transformed into gravitational radiation.

Of course, the difficulties associated with finding and fusing black holes would be quite formidable. First, we would need to seek out existing black holes; creating artificial black holes would require so much energy that it's unlikely there would be any payback. Fortunately, evidence seems to indicate that there should be plenty of black holes near the center of the galaxy.

Once these collapsed stars were found, they would need to be moved, a process that would require extreme care, since falling into the event horizon of any of them would be lethal (as the unlucky astronaut story, in Chapter 3, revealed). Most likely, extremely powerful electromagnetic fields would be used to steer the black holes—that is, assuming that these bodies are electrically charged. Electrically neutral black holes would be much harder to manipulate, since they wouldn't respond to the tugging power of external electromagnetic forces.

Finally, after these massive objects are fused, a method would be required to collect the consequent radiation. Huge "solar" panels of energy-storing material could be set up in a spherical region just outside the impact zone. Once filled with power, these batteries could be transported away from the collision region via wormhole tunnels. Or, even simpler, wormholes themselves could be used to transfer the energy directly away from the area. One would need to place one set of mouths of these wormholes close to the impact region and the other set closer to a populated area of the universe. Then, after the black holes collided, the wormholes would transport the enormous postimpact energy output away from the danger zone and out to a safer area. There, power-transmitting devices could be placed to carry the power back to where it was most vitally needed.

A far easier way of extracting energy from black holes involves using them as direct matter-energy converters. If a material object is thrown toward the event horizon of a collapsed star, it begins to radiate away a significant portion of its mass as it draws near. For objects rapidly spiraling into the horizon, the efficiency of this process is just 5 percent; however, for more slowly moving chunks of matter, this efficiency can be increased to as much as 40 percent. Thus, more than one-third of the mass of a piece of debris could be directly converted into energy. This is an extraordinarily high figure, much higher than the efficiency of nuclear fusion.

All that is needed for this scheme to work are a black hole, a constant source of debris, a way of transporting the matter into the event horizon, and a means of collecting the consequent energy. Once again, finding a black hole shouldn't be difficult near the galaxy's core. Supplying debris wouldn't be a problem at all: there's an infinite supply in the universe. In fact, updated versions of current "trash-to-steam" conversion processes, operations in which garbage is burned for fuel, could involve tossing all the earth's rubbish into black hole "waste bins" and collecting the energy produced.

How could the matter be transported to the black hole and the energy siphoned back to earth? Wormhole tunnels are obvious candidates for these procedures. These gateways could be used for both purposes, taking debris to the event horizons and scooping up the consequent power produced by the matter's radiation. All of these operations would be quite safe; human crews could remain far out of reach of the gravitational vortex.

Strong evidence that the process of dumping matter into black holes can yield a tremendous amount of energy comes to us in the form of recent information concerning the extremely bright, extraordinarily distant celestial objects known as quasars. Recall that white holes were once considered prime candidates for the driving engines of quasars until it was discovered that they would be too unstable to do the job.

Quasars are fantastically powerful emitters of energy, far more luminous than a hundred galaxies. Yet the power source driving them is extremely small by cosmic standards: barely one light-year across, only a thousand times greater than the diameter of the solar system. Compare this, for example, to the width of our Milky Way galaxy: more than 100,000 light-years. Thus, although a quasar is much brighter than a typical galaxy, the engine driving its extreme brilliance is far smaller.

The discovery that the source of quasar power is so concentrated is a fairly recent one. Two Soviet scientists, Efremov and Sharov, and two Americans, Smith and Hoffleit, solved this problem indirectly. They measured the rates of variation of quasar radiation and found that they fluctuated roughly once a year. This means that the distance across a power source couldn't be more than a light-year; otherwise light would take longer than a year to cross it and the quasar's signal would take more than a year to vary.

Since a quasar is relatively tiny, considering the fact that it is brighter than a trillion suns, scientists were, at first, bewildered by this phenomenon. Gradually it dawned on many scientists that a simple solution was possible to this dilemma: quasars are extremely massive black holes in the process of gobbling up matter and spitting out radiation. Evidence exists that all galaxies possess such giants in their nuclei, embodying churning dynamos of power; for quasars, though, this activity may be particularly strong. The mammoth black hole powering a quasar, according to this theory, has a mass equal to hundreds of millions of solar masses and a diameter about the size of Jupiter's orbit around the sun. Matter, in the form of stars, gases, and other types of interstellar debris, pours into the hungry giant at a continuous rate. Radiation, in turn, spews out of the region as the material is consumed.

So, in the current scientific model of quasars and galaxies, these objects are simply extreme versions of the same phenomenon, namely the accretion (gobbling up) of matter by supermassive black holes. For distant quasars, the resulting radiance is far more pronounced, but even in nearby galaxies the same process of mass absorption by black hole giants can be clearly observed by astrophysicists. All galactic cores produce massive amounts of electromagnetic radiation; it is likely that in most cases black hole accretion is the source.

The verdict on the existence of supermassive black holes is not yet in. Most physicists now believe in them; however, absolute, irrefutable experimental proof has not yet been supplied. The Hubble space telescope, launched in 1990, and the Gamma Ray Observatory, carried into orbit by the *Atlantis* space shuttle in 1991, should soon provide much more information (to date, detailed analyses of telescopic reports haven't indicated any black holes). Energetic gamma rays, emitted by stars entering supermassive black holes, will, most likely, provide key proof of the existence of these objects. Thus, the Gamma Ray Observatory could be of tremendous help in understanding the fate of stars, galaxies, and, ultimately, ourselves.

One might easily assume that supermassive black holes are the last word in huge, but there may be even bigger black objects in the universe. In 1991, astronomers at the Mauna Kea Observatory discovered a dark, mysterious, and enormous body lurking within a shell of luminous gas. The size of this object, found while studying a bright galaxy named NGC 6240, is ten to one hundred times more massive than any black hole previously believed to be possible. Measurements indicate that this monstrous vortex is about as massive as our own galaxy but only one-ten-thousandth as big. Theorists believe that this dark object may be a dead or dormant quasar just beginning to feed on fresh matter from a nearby galaxy. Since this discovery is so recent, it remains to be seen whether it will provide the final proof that black holes exist.

As we have seen, there are many ways to extract energy from a black hole. Let's now add one more method to the list: gravitational superradiance. Superradiance, discovered by the Russian physicist Ya. B. Zel'dovich, works by shining electromagnetic or gravitational radiation into a black hole and collecting the scattered waves. If this is done exactly right, the intensity, and hence the power, of the outgoing radiation is greater than that of the incoming waves.

Thus, in superradiance a black hole acts as a sort of cosmic amplifier. Like the speakers of a stereo set, it takes incoming signals and magnifies their strength, adding energy in the process. Since the extra energy comes from the rotational power reserve of the stellar relic, only spinning black holes can exhibit this amplifying effect.

The efficiency of superradiance as an energy source is not very high; for incoming electromagnetic waves the amplification factor is only 4.4 percent. This corresponds to just a small transfer of energy from the black hole to the outgoing waves. However, this effect can be augmented to 138 percent amplification by use of little-understood gravitational waves.

Even with electromagnetic waves as a source, it may be

possible to generate a tremendous amount of power by use of superradiance coupled with a clever magnification scheme involving a large reflective surface. Suppose a rotating black hole is completely surrounded by a giant spherical mirror that reflects all electromagnetic radiation. Then, electromagnetic waves are sent into the black hole, amplified by the superradiance effect, and scattered out to the mirror. The mirror, in turn, reflects the waves to the collapsed star, where they are reamplified. After a series of bounces between the black hole and mirror, the intensity of the wave is amplified by a considerable amount.

This process cannot continue forever. The intensity amplification builds and builds with each reflection, creating more and more radiation pressure, until eventually there is a huge explosion. Thus, Zel'dovich's superradiance scheme is a blueprint for a sort of gravitational "bomb" far more powerful than any nuclear explosive.

As we discussed in Chapter 3, black holes leak; according to Hawking's famous results, they typically radiate away all of their energy in roughly 10^{66} years. Now we see that this slow trickle can be enormously magnified by use of the proper techniques. By using ballistics, fusion, accretion, or superradiance, a substantial portion of black hole energy can be drained and used. Rather than taking astronomical lengths of time to produce results, these processes can take place over far more workable time scales. In the cases of superradiant amplification devices and gravitational bombs, the results might even be surprisingly brief, perhaps even close to instantaneous. Clearly, black holes are volatile creatures: energetic, unstable, and often deadly!

*B*urrowing with Wormholes

Black holes are cosmic pariahs. No one with any sense would dare to enter the swirling, caustic stomachs of these celestial leviathans. It would seem that their interiors are forever off limits to exploration.

Fortunately there exist cosmological endoscopes to probe the mysterious innards of these creatures. The "father of white hole theory," Igor Novikov, along with his colleague, Valery Frolov, has devised a clever method of exploring black hole interiors by use of traversable wormhole gateways. Using their technique, safely venturing into black holes may someday be possible.

Novikov, one of the leading Russian black hole and wormhole theorists, grew up during the brutal Stalin era. When Novikov was a youth, the Soviet Union itself might as well have resided in a black hole; emigration was impossible for all but a privileged few, generally those of the Communist party cadre. As a child, Novikov endured the terrible hardships of World War II and survived the postwar famine. Several years after the war, he enrolled as a physics student at Moscow University.

While engaged in postgraduate studies, Novikov found out the hard way how demanding theoretical work in general relativity could be. His thesis adviser, Professor Zelmanov, assigned him an extremely tedious and complex calculation in this field. Denied a request for an assistant, Novikov set out with pencil and paper (pocket calculators had not yet been invented) to solve what would be his dissertation problem. In the end, Novikov won the endurance contest; it took him hundreds of hours, but he completed the task. This strenuous training prepared him for a lifelong series of arduous calculations involving stars, galaxies, white holes, wormholes, and other celestial objects.

After receiving his doctoral diploma, Novikov began a long-term collaboration with Zel'dovich. He also began to engage in a highly productive dialogue with Kip Thorne. Recently, Thorne and Novikov have combined forces to delve into some of the more perplexing issues involving traversable wormholes. Now, at the Space Research Institute in Moscow, Novikov works closely with his old friend from Caltech in examining questions involving the use of wormholes as time machines (we'll discuss this in the next chapter). Together, Thorne and Novikov hope to iron out the remaining issues concerning wormhole space travel.

Fortunately, the end of the cold war has meant that Soviet-American and Soviet-European collaborations of this sort are far less problematic than they used to be (though the recent collapse of the Soviet government has made it extremely difficult for scientists there to obtain funding). In a scientific exchange that would have been virtually impossible during the worst years of the cold war, Novikov's co-worker, Frolov, has spent a considerable amount of time at the Theoretical Physics Institute of the University of Minnesota. Together, Frolov and Novikov have recently published an intriguing research paper that attempts to examine how black holes and wormholes would interact. They imagine what would happen if a wormhole gateway were located such that one of its mouths was held in place, by an external force, very close to the event horizon of a black hole. The second mouth would be located at a great distance from the black hole.

The authors show that such an arrangement would be highly conducive to black hole mining. The radiation cloud surrounding the black hole, in the turbulent region near its event horizon, would be an excellent source of energetic material (because of some of the processes we have already described). This fuel could freely travel though the throat of the wormhole and enable the second mouth to serve as a font of energy.

Imagine, however, what would happen if the external force holding the first mouth in place were to fail. Then this wormhole entranceway would plunge rapidly into the dark abyss, falling without hindrance through the event horizon and ending up in the mysterious region within. In doing so, the black hole and wormhole would form a curious amalgam. The resulting entity would be neither a gateway nor a dead end street, but something strangely in between these two extremes.

Having a wormhole terminus inside it would open up the black hole interior to inspection and exploration. Brave voyagers could enter the black hole through the event horizon and escape through the wormhole's throat. Scientists could carefully study the collapsed star's unusual gravitational effects. Since wormhole connections offer the hope of rescue missions, the risks associated with black holes would be tremendously reduced, rendering these collapsed objects far less threatening.

Suppose an errant wanderer does, in fact, get trapped inside one of these cosmic maelstroms. Ordinarily, black holes would offer absolutely no means of escape. Solitary black holes, without wormhole tunnels, provide nothing but solitary confinement and, ultimately, a terrible demise for those unlucky enough to enter them.

Fortunately, black hole rescue missions would often be possible by use of wormhole connections. An astronaut who inadvertently passed through the one-way entranceway of a black hole would have at least some hope of being saved. There are, however, a few important preconditions for such a bold attempt at liberation. First, the black hole must be large enough that there is a great enough duration between the astronaut's entrance into the horizon and his or her ultimate plunge into the central singularity. Otherwise, a rescue attempt would be futile: there wouldn't be enough time for success. Second, the rescuers must detect the wayward astronaut themselves—they can't just sit and wait for a distress signal. A signal from the astronaut would be impossible, because nothing can escape a black hole, not even radio waves. Third, they must act right away: to do otherwise would jeopardize the whole mission.

With these provisos in mind, let's imagine how a rescue mission would work. In the first place, the mission commander would have to be alerted that someone had dropped into the abyss. The commander would then use guidance controls to steer the mouth of a wormhole such that it dropped into the event horizon at the right spot. This steering could be done via powerful electromagnetic or gravitational fields which would latch on to the mouth and drag it to a point near the horizon. There the mouth would sink into the black hole and "land" at a point near the distressed individual. Finally, the person to be rescued would steer his or her ship into the wormhole mouth, pass through its throat, and then emerge in deep space, where he or she could then set sail for home.

With wormhole connections unveiling their inner depths, black holes aren't really so black after all! In fact, using the methods described by Frolov and Novikov, material can be transferred from the interiors of these objects to the outside world, yielding yet another means of extracting energy from black holes. All that is needed for this form of black hole mining is a wormhole situated such that one of its entrance-ways lies within the event horizon. Then energy can be drained out from the central portion of the collapsed star, into the wormhole's mouth, in through the throat of the wormhole, and finally out from the second mouth. This second mouth can be moved to any part of the universe, creating an energy pipeline to any place where the black hole's fuel is vitally needed. If, for instance, it is placed near earth, then a direct energy connection between the black hole and earth is possible, yielding a tremendous source of power.

Unlimited Power

Unrestricted freedom to travel, unlimited power over nature, and unfettered control over destiny are three of humanity's ultimate fantasies. Wormholes put these long-held dreams within reach; they help span seemingly unbridgeable gaps, tap into seemingly impenetrable power sources, and change seemingly inalterable circumstances (such as plunging into black hole event horizons). Unfortunately wormhole gateways are still only theoretical constructions; yet, given humankind's extraordinary creativity and scientific ingenuity, it is entirely possible that these inventions will be available for use within the next few centuries. Then, travel across the universe will become as commonplace as transatlantic voyages are today. Tremendous energies will be harnessed to create a new society of unbelievable affluence and leisure. And finally, as we shall see in the next chapter, time and destiny themselves will become targets for humankind's controlling grasp.

If wormholes someday become the veins and arteries of a new transgalactic civilization, black holes will then provide the lifeblood. With their tremendous capacities for collecting and storing energy and their almost certain abundance in the universe, these cosmic cauldrons represent an enormous source of precious fuel. All that would be needed for their successful exploitation as an energy reserve is a means of mining them. In this chapter, we have mentioned several possible ways, at least in theory, to mine black holes. These include dredging the sea of energy outside the event horizon, slowing the black hole's rotation with a ballistic approach and capturing its spin energy, tossing material into the black hole and collecting the resulting radiation, using the black hole's superradiant amplification of incoming waves, and, with Frolov and Novikov's technique, dropping one entranceway of a wormhole into the event horizon. If we are able to use one or more of these methods, it is entirely likely that black holes will become the next millennium's virtually inexhaustible fuel supply.

Perhaps by the year 2500, or even sooner, black hole mining expeditions will become commonplace throughout the galaxy. Armed with maps of all of the wormhole connections throughout the cosmos, prospectors will set out into the vastness of space in search of the fuel needed for the technological development of future civilization and for the growth and dissemination of the human population. To find and retrieve their valuable cargo, these hardy souls will venture forth into an unimaginable kingdom—a world that lies perpetually shrouded from view, a topsy-turvy realm of immense gravitation—in short, the lair of the black hole.

CHAPTER 7

THE FUTURE WAS YESTERDAY

*E*scape from the Time Stream

Someday, with the help of cosmic wormholes or other interstellar gateways, we will breach the barriers that now prevent us from engaging in rapid travel through space. But what of that other straitjacket that thwarts our efforts at full understanding of the universe: the hidden constraint posed by time itself? The "force" of time carries us with it on a one-way journey along the fourth dimension of the universe. Although free motion is possible along spatial paths, strangely enough, it's not possible along the temporal. Yet Einstein's theory of relativity seems to suggest a rough *equivalence* between the spatial and temporal modes; surely if unconstrained space travel is achievable, unfettered time travel ought to be possible as well. Thus it is likely that the spatial transport technologies produced during the coming age of interstellar spaceflight will also enable free temporal journeys to take place.

The possibility of traveling through time to the past or the future is one of humankind's longest held dreams. We ponder historical events and wonder what would happen if they were altered by someone stepping back in time from the present, equipped with vital information about these events. For example, it is interesting to consider how the chaos that followed the American Civil War would have been affected by the prevention of Lincoln's assassination. If a time traveler, armed with knowledge of the circumstances of the murder, physically stopped John Wilkes Booth from entering the Ford Theater, might the continued leadership of Lincoln have served as a way of healing the wounds of the Civil War much more rapidly?

What if the murders of Martin Luther King, Jr., the Kennedy brothers, John Lennon, and Olaf Palme (the Swedish statesman and peace activist) had been prevented by time travelers? Might the world today be a more peaceful place? Or suppose Hitler, Stalin, Idi Amin, and Pol Pot were removed or eliminated before their reigns of terror began. Perhaps millions of lives would have been saved, including those of brilliant poets, inventors, and other creative thinkers, and our civilization today would be all the richer. Might a precise and well-planned alteration of history lead to a saner and more prosperous society?

Given the natural public fascination with these issues, it is not surprising that a number of short stories, novels, television programs, and films have dealt with the ramifications of voyaging through time. Stories about intrepid explorers stepping out of the time stream into earth's past or future have continued to baffle and entertain us. Paralleling the rise of modern technology, which has put hundreds of miraculous inventions into our living rooms, the public interest in time machines has continued to expand.

Although the time travel story genre has blossomed in popularity only in the past century, these sorts of tales have existed almost as long as the American nation. One of the oldest American stories, *Rip Van Winkle* by Washington Irving, is an account of an unfortunate journey through time. In this famous tale, Van Winkle falls asleep and wakes up a number of years later, only to find that all of his friends and family members have aged beyond immediate recognition and then to realize to his horror that he cannot return to his youth. Another popular work, *Looking Backward* by Edward Bellamy, is a far more uplifting story of a Boston man who falls asleep in a soundproof vault and wakes up more than one hundred years later in a utopian socialist society. In *A Connecticut Yankee in King Arthur's Court* by

Mark Twain, another New Englander is transported to Arthurian England, in this case by a blow to the head.

In each of these early examples, the means of temporal transport is nonmechanical; rather, physical methods such as sleep or unconsciousness are used to enable the temporal shift to take place. These methods are similar to the cryonic suspension techniques described in Chapter 1. Dormant explorers don't actually leave the time stream; instead, they lose consciousness for a lengthy interval and are merely unaware of time's passage.

A time machine, in contrast, would enable wayfarers consciously to step out of the river of time and to reenter it at the moment of their choice. Hundreds of science fiction stories have detailed journeys of this sort using devices ranging from modified bicycles and automobiles to police phone boxes and specially constructed vehicles. In each story, a method is postulated by which the links between human time and physical (clock) time can be temporarily severed.

The first story about a time travel device was a tale published in 1881 by Edward Page Mitchell, an editor of the New York Sun. His fictional narrative, entitled "The Clock That Went Backward," was virtually unknown until it was reprinted in 1973.

It was not until the publication in 1895 of H. G. Wells's novella The Time Machine that the time travel genre was launched as a theme of science fiction. This short novel, a highly original look at what might happen if a man were to journey into the far future and safely return to report about it, drew its author much acclaim. Later made into two feature length films, it sparked the public imagination in an exceptional manner.

Wells's story begins with an after-dinner discussion about the nature of space and time. The leader of the conversation, a man identified only as the "Time Traveller," remarks that spatial distances and temporal durations should be considered as some of the components of the same four-dimensional entity. Three of these dimensions compose the three directions of spatial motion, while the fourth is time. These components are physically indistinguishable and can only be differentiated by the fact that humans perceive time as being unidirectional. As the Time Traveller explains: "There is no difference between time and any of the three dimensions of space except that our consciousness moves along it. . . . Scientific people know very well . . . that time is only a kind of space."

The Time Traveller then proceeds to show how motion in time can be achieved by use of a properly designed machine. Then he surprises his dinner guests by revealing that he has, in fact, built such a contraption. He boards this device and travels forward in time, finding the future society to be radically different from his own. Returning to the nineteenth century, merely a few minutes after he had left, he brings proof that control of time is indeed possible.

Time machine stories have become a staple of twentieth-century science fiction. Countless writers, from Ray Bradbury to Robert Heinlein, have written tales of this sort, exploring the implications of this unusual type of travel. For instance, in *The End of Eternity* by Isaac Asimov a future civilization develops a means of exploring the ages through a sort of "kettle." By programming the controls on this device, one can transport oneself to the century of one's choosing. An elite group, called the Eternals, commands this machine and uses it to maintain control of all of history and to tamper with historical events to suit their purposes. Eventually their society is destroyed by a historical intervention that deliberately eliminates the Eternals organization before it has been founded.

In Ian Watson's story "The Very Slow Time Machine" scientists observe a man who acts as if his personal clock were reversed in directionality—running backward in time, that is. It turns out that he is from the future but is moving back in time in order to "catapult" himself even further into the future. His time machine is set up so that one must experience a short amount of time reversal in order to fling oneself ahead in time; in other words, there is a sort of slingshot effect for time travel.

Gordon Dickson in his novel *Time Storm* paints a picture of a catastrophic disruption of the continuity of the earth's time frame in a sort of "time quake." Rifts and faults are opened up in the temporal strata so that moments in history are experienced at different times in different places. Thus cavemen exist in one part of the earth while contemporary people live in another sector and aliens from the future dwell in yet a third region. By moving from one place to another one can travel in time; hence, ordinary vehicles double as time machines.

In a very recent tale, "The Toynbee Convector," written by Ray Bradbury and published in 1988, a very clever time machine hoax is concocted. A man living in the year 1985 falsely claims to have returned by time machine from a peaceful, prosperous future. Producing strong documentary evidence to support his claim, he influences society in a profound way: a spirit of optimism infects the culture and all citizens unite to create a utopia. In this manner the man's dream becomes a self-fulfilling prophecy: the ideal world of the future is constructed by those of the present. When they realize the falseness of his claim, it is too late; the world has already been favorably transformed beyond recognition.

A number of popular television programs and movies have echoed these themes. In a 1960s television series, *Time Tunnel*, unlucky victims of an experiment gone awry are flung from one period of history to another. In each era they witness pivotal historical events but are powerless to change them (the more recent series *Quantum Leap* is a variation of this premise). The most popular British science fiction television series of all time, *Dr. Who*, is built on the premise that the protagonist can journey through time (and space) at will via a specially designed police call box. In the highly successful television and movie series *Star Trek* time travel is a recurrent plot device. In one of the television episodes of this series, for instance, members of the cast encounter a library with a massive selection of historical videotapes. By choosing a tape from this collection and walking through a special portal they can travel into any period of history that they would like to explore.

Finally, in the very recent humorous films *Bill and Ted's Excellent Adventure* and *Back to the Future* young boys encounter mad scientists who whisk them away from the present in thrilling adventures through time. They experience firsthand the incongruities involved when people from two different eras encounter each other and have to deal with radically different customs and mores.

Time and Relativity

Clearly the prospect of being able to travel at will into the past or future is exciting. The question is whether it is realizable. Obviously, we each travel into the future at a rate of twenty-four hours per day. However, this mundane sort of time travel hardly satisfies our more grandiose aspirations. We would like either to speed up the clock or to reverse it —in other words, to discard the constricting straitjacket of temporal uniformity.

Amazingly, time travel does indeed appear possible through specially designed traversable wormholes if, in fact, methods of wormhole construction are developed. Morris, Thorne, and their co-worker Ulvi Yurtsever have recently devised a theoretical means of using wormhole gateways for temporal voyages, a way in which the past or future could be reached by star ship flight. Before we discuss their model, however, let's briefly examine other proposed scientific methods for journeys through time.

Modern physics provides us with many schemes for time travel, ways in which durations can be distorted and simultaneous events can be rendered asynchronous. These ideas derive from Einstein's work in relativity and continue to be revised as new information becomes available.

We have seen in Chapter 2 that special relativity provides a simple means of time dilation: that is, traveling near the speed of light slows down the clocks of objects relative to those of a fixed observer. Thus, this remarkable theory suggests that journeys into the future are distinct possibilities, given the technical knowledge of high-speed space travel. If one were interested in traveling forward in time, one would "merely" need to hop on a fast enough spaceship and return after a sufficient lapse in earth time.

How would future-directed time travel work in practice? For one thing, it could give rise to some interesting schemes to pilfer money. Imagine, for instance, what would happen if, in the future, a thief were to deposit money in an account, steal a spaceship, then use the vehicle to travel forward in time to collect the interest on the funds. This scenario would be entirely possible if near-light-speed travel were to exist. The thief could place one thousand dollars in a high-security "no questions asked" bank account in Switzerland, where it would accrue 10 percent interest compounded annually, let's say. Then he could use his cunning to break into a spaceport and steal one of the vessels. Piloting the pirated craft, he could then set the speed controls at 99.99 percent of the speed of light and zoom off into space. He could stay in the ship for three years of his ship's (and own) clock time before returning to earth.

Landing back on earth, the thief would find that enormous changes had occurred. Although he would feel and appear only three years older, he would find that two hundred years of earth time has passed since he left. Nobody he would have known before he set out on his journey would still be alive. Second, he would be extremely wealthy from all the interest accrued in his bank account. The $1000 that he placed there would have grown to almost $200 billion. For the mere cost of three years of his life spent in space, his profit would be almost 200 million times his original investment.

This example illustrates one of many possibilities for future-directed time travel if near-light-speed space travel were perfected. Any sort of activity that would involve hastening the arrival of the future would indeed be possible for an explorer. Special relativity ensures that altering the speed of one's own internal clock relative to an external clock would be, in theory, readily achievable via extremely rapid transport. Practically, of course, one would need to find some way of physically enabling spaceships to fly so quickly, but theoretically no scientific barriers would hinder efforts to do this.

Past-directed time travel, however, poses far more problems. Although special relativity allows journeys into the future for objects traveling close to the speed of light, it seems to preclude journeys into the past. To use the time dilation relativistic technique to reverse the direction of an observer's experiences requires journeys faster than the speed of light.

That is, to travel into the past one would need to break the speed of light barrier mandated by special relativity. In the words of a famous limerick (by A. H. Reginald Buller):

> There was a young lady named Bright
> Who traveled much faster than light
>   She started one day
>   In the relative way
> And returned on the previous night.

Ordinarily, it is considered impossible for objects to travel at speeds faster than that of light. Why is this the case? If a substance could move faster than light, then it could be involved in an interaction that violates *causality*, the principle that cause precedes effect.

To see why faster-than-light travel would be problematic, imagine what might happen if a frisbee were created with such a property. When this frisbee was released, it would soar through the air at ten times the speed of light.

Now, picture what would happen in a frisbee game between two players, Sandy and Mandy, who were situated one hundred yards away from each other. Consider this account of the game: Sandy throws the frisbee to Mandy, who catches it an instant later. Naturally, from this description, it's clear that Sandy's action would be the cause of the frisbee's flight and that Mandy's would be the effect.

However, consider how this situation would *appear* to Mandy. Since the frisbee traveled faster than light, light signals bounced off it would reach her more slowly than the frisbee itself. So, first she would physically encounter the frisbee; that is, she would catch it. Then, she would see the frisbee directly in front of her, right before it hit her. Next, the frisbee would appear twenty-five yards away, then fifty yards, then seventy-five yards. Finally, she would view the frisbee's being thrown by Sandy. Thus the order of events, according to Mandy, would appear reversed in time. The principle of causality would seem to be violated, as cause would follow effect.

Which sequence of events would be the true one: the one perceived by Mandy or the one dictated by reason? This is a true paradox, engendered by the nature of light and communication. Anything that surpassed in speed its own image would naturally reach any particular destination before its signal. Thus signals from the past would continue to arrive long after the object itself and a picture would be painted of backward motion in time. While from the perspective of the object time would run forward, all external signals would lead to the opposite conclusion: reality and its image would stand in blatant contradiction.

Why can't one simply assume that the backward-in-time picture was the true one? Why not accept the visual illusion of reversed motion as the correct representation of reality—the frisbee's having traveled, in other words, from Mandy to Sandy?

A simple thought exercise shows that this clearly would not be the case. Suppose Sandy were to autograph the frisbee before she threw it. Clearly then it would be Sandy, not Mandy, who held it first; both women would agree that the autograph belonged to Sandy, not Mandy. Thus it is manifestly clear that Sandy would be the cause of the frisbee throw and Mandy's catching the frisbee would be the effect. Therefore, our paradoxical conclusion is confirmed: namely, effect (receipt of the autographed frisbee) *appears* to have preceded cause but actually must have followed cause.

The contradictions embodied in faster-than-light communication are thrown into sharp relief when one considers the nature of *tachyons,* hypothetical particles that travel at supraluminal (greater than the speed of light) speeds. During the 1960s there was considerable interest, both theoretical and experimental, in detecting such particles. Since nothing in relativistic physics precludes such objects, scientists thought there was a reasonable chance that they existed. Unfortunately, experiment after experiment produced null results: no such faster-than-light particles were found.

Theoretically, problems were also found with tachyons. A pivotal research article, "The Tachyonic Antitelephone," by G. Benford, D. Book, and W. Newcomb of the University of California at Livermore, showed that the use of tachyons for communication would lead to causality violations and logical contradictions. Benford, Book, and Newcomb made use of a 1917 statement known as Tolman's paradox, which proves that the existence of faster-than-light signals would imply communication with the past. That is, it would comprise a sort of "antitelephone." By use of such a device, one caller could send a message to another, directed into the second caller's past.

In their article, Benford, Book, and Newcomb provide an example of how such an antitelephone could be used, clearly depicting the philosophical problems involved. They imagine that William Shakespeare and Francis Bacon are engaged in a dialogue by antitelephone. The drama (or should we say antidrama) begins with Shakespeare's scribing a first draft of *Hamlet*. Wishing to have a kindred soul read his prose, the Bard then uses a tachyon transmitter to send the manuscript to Bacon. Naturally, since it is being sent by antitelephone, Bacon receives the transmission at an earlier time and can read the play even before it's been written! On this basis, he claims to be the author of the work.

Who then is the true creator of the play, Shakespeare or Bacon? On the one hand, it is true that Bacon possesses the original copy of *Hamlet* before Shakespeare has scribed it. On the other hand, it's absolutely clear that Shakespeare has written *Hamlet*: the work reflects his unique style and idiosyncrasies and is even in his handwriting. Thus, as in the frisbee example presented earlier, the cause of *Hamlet*'s creation (Shakespeare's writing it) occurs later than the effect (Bacon's reading it). Strangely, therefore, cause and effect appear here to outside observers in reverse temporal order. This is a most puzzling dilemma.

An even more serious problem arises when one considers the backward-in-time transmission of several messages, each contingent on the other. For example, suppose Abe and Gabe, two broadcasters, each possess antitelephones that can send messages back one hour in time into the past. They enter into the following strict agreement: first, Abe will send out a message at three o'clock if and only if he does *not* receive one at one o'clock; second, Gabe will send out a message right after two o'clock if and only if he receives one from Abe at two o'clock. The paradox is: does Gabe send a message or not?

The paradox is clear: the exchange of messages will take place if and only if it doesn't take place. This sort of irresolvable causal contradiction suggests that the use of tachyons for communication is highly problematic at best.

Because of the paradoxes embodied in faster-than-light travel, it is not clear whether or not time travel or backward-in-time communication is possible using this method. Other techniques for time travel, making use of ideas in Einstein's general theory of relativity, which seem to prevent some of the theoretical problems embodied in the idea of tachyons, have been proposed. These include methods involving cosmological models proposed by Gödel, Tipler, and Gott as well as black hole and wormhole schemes.

Esteemed as one of the greatest mathematicians of the twentieth century, Kurt Gödel is best known for a famous theorem stating that all logical systems are either incomplete or inconsistent. However, Gödel's work extended to many other fields of endeavor, including cosmology. In 1949, after an extensive study of Einstein's equations of general relativity, he found, as a solution, a set of universe models that exhibit rotation. Since astronomical measurements have shown that it is unlikely that our own universe rotates, it is unclear exactly what Gödel's models describe. Nevertheless, because Einstein's theory is so successful, any of its solutions must be seriously examined as a contribution to our further understanding of cosmology.

What is remarkable about Gödel's solution is what its existence implies about the nature of time travel. In contrast to the nonrotating Friedmann-Robertson-Walker (FRW) model most commonly used to describe the big bang theory of the cosmos, Gödel's rotating universe model implies that one can travel to any point in the world's past or future. Thus, the rotating cosmological model describes a sort of "universal time machine" in which the cosmos itself acts as a device to transport individuals from one era to another. Space travel in Gödel's model can lead naturally to time travel without violation of special or general relativity.

To experience a trip through time, in Gödel's universe model, one would need only to travel in a large circular path perpendicular to the model's rotation axis. The radius of this path must be greater than a certain minimum quantity that depends on the rotation rate of the universe. Along this route, the rotation of matter in the universe causes the directions of space and time to be twisted, so that time travel becomes as realizable as ordinary space travel. By continuing in a cyclical motion around the axis of this model, a simple voyage through space would become a journey into the voyager's past or future as well.

For Gödel's model the rate of rotation must exceed a certain minimum amount for temporal voyages to be possible. This speed is determined by the fact that the centrifugal force of the universe must be great enough to balance the gravitational attraction of the matter contained within it. Applying current estimates of the density of the physical universe to this model yields a period of rotation of 70 billion years —the universe makes one complete turn during this time period.

For a universe spinning at this rate, the circular distance required to travel to one's own past or future would be roughly 100 billion light-years. This is a tremendous distance: a million times greater than the breadth of the Milky Way galaxy. In fact, according to detailed calculations, at this rate of rotation our universe would be too small for such journeys even to be possible, let alone manageable. To reduce the distance needed for time travel, the universe would have to spin at a much, much faster speed.

Does our own universe experience the rotation required for Gödel's time displacement method? Apparently not— rotation this great would produce detectable effects, none of which has been measured. In particular, if the cosmos were spinning, there would be an *anisotropy* (directionally skewed distribution) in the background radiation left over from the big bang: in other words, the light remaining from the initial explosion of the universe would appear of varied intensity in different directions. Since this anisotropy hasn't been observed, it's apparent that there isn't a significant amount of rotation.

Since our own universe does not appear to spin, it is unclear what physical significance can be assigned to Gödel's rotating model. On the other hand, a more recent time machine scheme proposed by the physicist Frank Tipler involves possible structures within our own universe and hence has more direct consequence.

Tipler's proposal involves a massive infinite cylinder, rotating such that its surface travels faster than half the speed of light. The effects of such an object, following Einstein's general relativistic prescription that mass distorts space-time, would be to warp the region around it. Near the cylinder, space and time would be extremely twisted, much like batter near the rotating spindle of a blender. In this region time's axis would be so altered that astronauts could travel into their past or future merely by orbiting the cylinder in a clockwise or counterclockwise manner.

Suppose Tipler's cylinder were constructed (this is purely conjecture, since it is impossible to create something infinite). How could it be practically applied for time travel? Let's say Tipler wanted to meet Einstein to discuss mutual cosmological interests. Since Einstein died in 1955, Tipler would need to use some means of temporal transport, such as his own device, to accomplish this rendezvous. He could, for instance, board a spaceship, blast off into space toward the rotating cylinder, reach the cylinder, circle it several times in the direction of its spin, then return to earth. Arriving back on earth, he could then meet Professor Einstein and explain his remarkable time machine to him.

Unfortunately, the prospects for finding or building Tipler's cylindrical time machine are somewhat poor. Clearly a strictly *infinite* device is out of the question: it would involve an infinite amount of material. On the other hand, as Tipler later found, a *finite* cylinder would be extremely unstable, and thus unsuitable for temporal voyages. Regardless of whether or not Tipler's machine could be constructed in reality, the science fiction writer Poul Anderson has already appropriated it for use as a plot device in his novel *The Avatar.*

In 1991, the physicist J. Richard Gott of Princeton proposed using cosmic strings for time travel. *Cosmic strings* are the hypothetical filamentlike remnants of vast defects in the unified fields that formed the universe. According to theory, these strings were created as these fields "broke up" into the present-day forces and particles of nature. Thus, cosmic strings represent "cracks" in the bedrock of space, just as fault lines indicate defects in the substructure of the earth's surface. Supposedly, they were "frozen" into the universe's fabric as the rest of the cosmos cooled down from its initial fireball.

If cosmic strings were to exist today, they would be incredibly massive in spite of their thinness. Their density, exceeding a thousand trillion tons per inch of length, would cause distortions in the paths of light rays in their vicinity. This effect, a direct consequence of the general relativistic prediction that massive objects curve space-time, would lead to a resulting alteration in the flow of time in the region.

In Gott's journal article, published in the prestigious *Physical Review Letters*, he describes two straight cosmic strings that do not intersect, moving in opposite directions at speeds extremely close to that of light. Gott shows, by detailed calculations, that observers could visit their own past simply by traveling in a giant orbit around the pair of strings. The set of cosmic strings would act to provide an enormous temporal "boomerang," propelling circling spaceships back in time.

Gott's proposal is superior to Tipler's in that cosmic strings might indeed be found in nature, whereas infinite rotating cylinders are merely mathematical abstractions. However, cosmic strings haven't yet been experimentally detected. Thus the cosmic string model is even more speculative than current theories regarding black holes and quasars.

In containing the possibility of backward time travel, Gödel's, Tipler's, and Gott's models all allow for causality violations in the form of *closed timelike curves* (CTCs). If a region of space contains a CTC, then communication with the past is possible from one time period to a previous one without violating the long-held prohibition against faster-than-light signals. Hence a CTC light ray can intersect itself by traveling into its own past and emerging at the same point where it was at an earlier time. We shall speak about these more in the next chapter.

*B*lack Holes and Time Travel

Unlike Gödel's, Tipler's, and Gott's fairly abstract models of travel through time, the idea of black hole time travel is somewhat more solidly based in observational astronomy. Although Gödel's universe probably doesn't describe our own, Tipler's cylinder is highly unstable, and Gott's model is based on cosmic strings that haven't yet been detected, black holes almost certainly exist and have, at present, a far more secure place in the world of astrophysics.

In the discussion that follows, we shall exclude the possible effects of the coupling of black holes with wormholes. As we saw in the previous chapter, once traversable wormholes enter the picture, models of black hole interiors become far more complicated. Later we'll discuss the promising attempts to construct time machine models based on traversable wormholes themselves.

We recall that the simplest sort of black hole can be described by the Schwarzschild model. When scientists first realized that this simple solution of the Einstein equations contained a sort of cosmological connection between two universes, the actual physical meaning of such a linkup was intensely debated. One of the questions that theorists pondered was whether or not time travel was possible by using the strange space-time geometry of this model.

Now that the Schwarzschild solution has been clearly identified with electrically neutral, nonrotating black holes, the issue can be framed as follows: could astronauts enter such a dark object through the one-way membrane of its event horizon, travel through time while inside the space-time inversion zone situated within the black hole's boundaries, then journey out again through another black hole, located in a second universe (or another part of our own universe)?

This question is somewhat subtle, but ultimately the answer is no. It's true that inside the event horizon of every black hole lies a space-time inversion region, where the spatial and temporal coordinates exchange properties. Technically speaking, there is a reversal of roles in the space-time metric: the spatial coordinates switch from real to imaginary numbers and the temporal, from imaginary to real quantities. When this occurs, time and space also can be said to change roles; it becomes possible to move freely in time, but not in space.

Imagine a voyager (call him Ishmael) who enters the one-way membrane of a Schwarzschild black hole. As described earlier, he feels no "bump" on entry; he simply notices that it is now impossible to turn his ship around. Once inside this celestial leviathan, he immediately enters the inversion zone: space and time remain switched for the rest of his journey to the black hole's central singularity. In this inverted region, as opposed to the "normal" cosmos, Ishmael can travel through time to his heart's content but is drawn inexorably through space in a unidirectional motion. He cannot prevent his ultimate descent into the crushing center, no matter how hard he tries. John Taylor, author of *Black Holes: The End of the Universe?*, has dubbed this region "the world of the topsy-turvy."

One might think that since free time travel is possible within the inversion zone, Ishmael might escape his terrible plunge to the center by either slowing down his temporal passage, hovering at about the same point in time, or maybe even reversing the hands of the clock. He might even wish to follow the time-stopping example of Martin, the protagonist of Robert Bloch's short story "That Hell-Bound Train."

In this tale, Martin, a drifter, makes a Faustian deal with the conductor of the train to Hell. Martin agrees to board the train after he dies, in return for an extraordinary watch that can stop time whenever he pleases. He decides that he will wait for a moment of perfect happiness and then use the watch to spend eternity in that blissful instant. Unfortunately, Martin dies before he can halt time and finds himself aboard the Hell-Bound Train. In the end, however, he beats the devil at his own game by stopping the watch while he is on the train. Thus, Martin and the motley crew aboard the vehicle manage to remain perpetually at the same moment in time and to forever avoid plunging into the fiery inferno.

Could Ishmael, on his flight through the black hole interior, stop the hands of time in a manner similar to Martin's? Could he arrange, by careful manipulation of the controls of his spaceship, to hover forever in the precise moment before he has to plunge into the darkened star's own fiery abyss? In other words, could he use the altered state of time in the inversion zone to save his own life?

Alas, it is impossible for Ishmael to rescue himself this way. While he is continuously plunging into the black hole's center, and perhaps even doing temporal somersaults in the process, his own personal clock, depicting the time rate of his consciousness, is ticking away in a ceaseless rhythm. In spite of being able to move effortlessly through the black hole's time line, his own fate is sealed from the moment he sets out into the sphere.

Moreover, Ishmael couldn't hope to escape from the black hole into the second connected universe. As we have discussed, the Schwarzschild connection between two universes is not traversable. No matter what, he must quickly meet his doom at the central singularity. Hence, since escape from a Schwarzschild black hole is impossible, attempting to use one as a time machine is futile.

Kerr black holes (see Chapter 3), with their distinctive rotational behavior, are somewhat better candidates for time machines. They possess several features that make survival prospects more hopeful, including ringlike and avoidable— rather than pointlike and unavoidable—central singularities as well as timelike, rather than spacelike, connections with other parts of the universe. As we recall, timelike links are far preferable since, unlike spacelike tunnels, they do not require faster-than-light travel, which would be problematic.

Rotating black holes can be divided into two categories: electrically neutral (ordinary Kerr) and electrically charged (Kerr-Newman). In both cases, there is a ringlike singularity, surrounded by a space-time inversion zone, which is further enclosed by a one-way membrane. Inside the ring, in each case, is a region which Taylor has dubbed "the no-man's land," in which time travel is theoretically achievable. Distinguishing the Kerr-Newman solution from the ordinary Kerr model is the fact that in the former case the no-man's-land doesn't extend beyond the ring, whereas in the latter example, this region also surrounds the ring in a thin envelope. Once inside this envelope, journeys into the past or future are possible by orbiting the black hole axis of rotation in the appropriate direction.

Suppose Ishmael's uncle Ahab, a very clever spaceship captain, wishes to try his luck by using a charged, rotating black hole as a time machine. First, he steers his ship into the one-way membrane of the dark object. He is extremely careful to enter the black hole in such a way that his ship is aimed away from the actual singularity and toward the envelope region surrounding the ring. Obviously, he doesn't want his craft to be pulled apart like taffy by the infinitely large gravitational tidal forces of the singularity.

After passing through the event horizon, Ahab enters the space-time inversion zone. There, he carefully guides his vessel until it follows a direct course into the ring envelope. Once he is inside the envelope, he notices the curious effect that space and time have switched roles, once again, back to their ordinary modes of behavior. As in the outside world, space and time behave normally within the narrow envelope engulfing the ring. Curiously, however, cyclical paths around the axis here result in temporal journeys.

Now safely within the ring envelope, Ahab begins his voyage through time. Each time he circumnavigates the region, depending on which way he goes and how fast the black hole spins, his clock is set back (or ahead) by several years. Here Ahab strictly controls the passing of the ages by deciding how many times he wishes to circle around the axis of rotation. When he is satisfied with his temporal voyage, he halts the ship's circular motion and veers away from the envelope and out through a second space-time inversion region. Finally, Ahab propels the ship back out into the normal universe, through the second terminus of the black hole (quite possibly a *white hole*), and, assuming that he can somehow escape, finds that he has traveled to a different part of space and a different period of history as well. After a lengthy journey, he returns to earth, disembarks, and begins a new life in this past or future era.

Technically speaking, the voyage that we have just described is attainable by careful use of rotating black holes.

However, as discussed in Chapter 3, Kerr black holes present a wide range of seemingly insoluble practical difficulties.

One possible solution to the many problems associated with entering Kerr black holes involves hypothetical celestial bodies known as *super-extreme Kerr objects* (SEKOs). A SEKO is a Kerr solution to the Einstein equations, which, unlike a rotating black hole, possesses the special property that its total *angular momentum* (momentum associated with spin) is greater than its total mass. This condition is highly difficult to achieve, but let's suppose, for the sake of discussion, that it is indeed possible to create a SEKO.

For such an unusual object, the safety problems associated with ordinary Kerr or Kerr-Newman objects are largely eliminated and time travel is therefore much easier. The reason for this fortunate arrangement is simple: for a SEKO, unlike a rotating black hole, the one-way membrane does not exist and the central ring singularity is thereby exposed. Thus free passage to and from the singularity is possible without running the risk of being unable to escape or of making an unwanted journey into another part of the universe. Furthermore, the ring envelope, the region for which time travel is possible by orbiting the axis of rotation, is so large that it covers the entire universe. Thus, if just one of these strange creatures existed in the cosmos, the entire universe would function as a time machine.

An astronaut can use a SEKO to travel through time with the minimum of discomfort. Instead of engaging in a perilous voyage through a hidden, black abyss, all he has to do is find out where the axis of the SEKO is located, then steer his ship in a circular motion about this imaginary line. By orbiting the SEKO axis several times, he can freely sample the past or future of any point in space without running any life-threatening risks.

The physical existence of a SEKO would create some rather disturbing philosophical dilemmas. Both black holes and SEKOs possess the curious property of embodying singularities, points where space and time seem to come to a halt. Close to these cosmic discontinuities, the laws of physics simply fail to function. The difference, however, between black hole and SEKO singularities lies in their exposure to the outside world: black holes are clothed, and SEKOs are naked. That is, within the confines of collapsed stars, the flawed and troublesome nature of these mathematical aberrations is sufficiently veiled as to make them less philosophically disturbing.

On the other hand, SEKOs possess so-called naked singularities. Like a bare ceiling with unsightly exposed pipes, the mathematical ugliness of such flawed creations is obvious for all the world to see. Lacking any shielding from the gaze of critical scientists, naked singularities, with their flagrant violations of physical theory, cause considerable aesthetic discomfort, to say the least.

Because of the practical difficulties in creating them and the philosophical problems in accepting the existence of the naked singularities within them, SEKOs generally aren't considered likely candidates for cosmic time machines. So, we can add them to the catalogue of theoretically possible but functionally irrelevant cosmological notions for time travel, a list which already includes faster-than-light theories; Gödel's, Tipler's, and Gott's models; as well as black hole methods. One far more promising approach to fulfilling these aspirations has not yet been discussed, however: using wormholes as time machines.

Wormholes as Time Machines

When physicists began to realize that black holes provided poor candidates for safe, efficient interstellar gateways, the solution to this dilemma became clear: traversable cosmic wormholes, specifically fashioned to allow rapid and secure passage through the heavens. Therefore, it comes as no surprise that theoretical astrophysicists have been hard at work developing better models for feasible time travel than those based on collapsed stars and that most of these schemes involve use of traversable wormholes.

The first proposal for use of wormholes as time machines occurred in an auxiliary section of Morris and Thorne's classic, original paper, almost tucked away as an afterthought. In this addendum, the authors comment on a scheme, using two traversable wormholes, for astronauts to travel back in time into their pasts.

In Morris and Thorne's scenario, they imagine two wormholes, each with two widely separated mouths, moving such that one set of mouths is traveling at a high-speed velocity with respect to the other set of mouths. For the sake of this discussion, call the first wormhole the "static" gateway and the second, the "moving" gateway, even though, according to special relativity, speeds of flight are all relative and what is fixed for one person is moving for another. Suppose further that the first mouth of the static wormhole is situated somewhat close to the second mouth of the moving wormhole, and vice versa.

Now picture that an astronaut wishes to use this set of wormholes to travel back into a past era of her life. She could do so by making two simple wormhole journeys, involving both the static and moving gateways. First she would enter one mouth of the static wormhole, travel through its throat, and emerge via the second mouth into another part of space. In other words, she would complete an ordinary interstellar voyage by wormhole of the sort described in Chapter 5.

According to her own clock time, corresponding to that of the static reference frame, she would surface from the second mouth a finite amount of time after she had plunged into the first (a year, let's say). However, using special relativity as applied to this situation, from the moving reference frame these two events (entry and exit) would be seen in reverse order. The reason for this is that the two mouths of the wormhole are sufficiently far apart that the separation between the two occurrences is considered spacelike in nature. In special relativity, spacelike connections can be seen in reverse order by observers situated in a frame traveling at a high speed relative to the linked events. Thus, according to the moving wormhole's reference frame, the astronaut would have left the second mouth *before* she entered the first (ten years earlier, let's say).

So if, after leaving the static gateway's second mouth, she then accelerated to the speed of the moving wormhole, she would travel ten years back in time according to her original clock. Finally, by entering the first mouth of the moving gateway and leaving it via its second mouth (one year after entrance into the first portal, let's say), she would find herself back at her original location, near the first mouth of the static wormhole. The only difference she would notice is that the calendar would read nine years earlier: she would have gained ten years by traveling through the static wormhole and have lost one year journeying through the moving one. Thus she would have effectively used the wormhole pair as a highly efficient time machine.

As soon as Morris and Thorne finished their classic article, including the account of wormhole time machines, they began to give considerable thought to the whole time travel question. Soon, what had been a peripheral part of their studies became pivotal. In particular, they began to devise a simpler method of time travel, this time using a *single* wormhole.

Joined by the Caltech researcher Ulvi Yurtsever, Morris and Thorne fashioned an incredibly straightforward means of temporal displacement. All that is required is a two-mouthed wormhole gateway, with one mouth traveling at a high-speed velocity relative to the second (call these the moving and static mouths). Following the special relativistic "time dilation" effect, clocks fixed to the moving mouth advance more slowly than those at the stationary mouth. Thus the time registered by the former devices is earlier than that measured by the latter.

In other words, the two mouths of the time machine wormhole enact exactly the same sort of behavior as that of the participants in the *twin paradox* thought experiment. In that famous exercise in special relativity, it is imagined that two twins separate; one remains on earth and the other boards a spaceship. The vehicle takes off, accelerates to near-light speed, heads toward a distant star, and then returns to earth. The relativistic behavior of the spaceship causes the twin on board to age much less rapidly than the one on earth. Similarly, clocks attached to the moving vehicle (more precisely, the *accelerating* vehicle) run much more slowly than those on the stationary object (earth).

For a brief time, it was thought that the twins experiment represented a genuine paradox. The supposed dilemma was that, since both the vehicle and the earth undergo rapid motion relative to each other, clocks situated at each reference frame should each slow with respect to the other as well, thus creating an impossible situation in which time dilation occurs for both parties. This, of course, presumes that the behavior of the spaceship with respect to the earth exactly mirrors that of the earth with respect to the spaceship. Therefore, observers on either reference frame must have exactly the same experience as those on the other frame and neither party can unilaterally age at a lesser rate than the other.

The solution to the twin "paradox" is quite simple. The situations for the moving and fixed observers in this thought experiment are emphatically *not* similar. The spaceship-situated twin undergoes acceleration, while the earthbound double does not accelerate at all. Hence, according to the predictions of special relativity, it can be mathematically proved that the moving observer does, in fact, age at a lesser rate than the fixed observer. There is no paradox here whatsoever: clocks on the spaceship do behave differently from those on the earth.

The Caltech wormhole time machine model borrows the results of the twins thought experiment to create a way of traveling into the past. In Morris, Thorne, and Yurtsever's innovative approach, the two mouths of the wormhole exhibit exactly the same motion as the twins in the experiment. The moving mouth accelerates away from the fixed mouth, then backs toward it again, causing clocks and observers in its vicinity to "age" less rapidly than those in the area of the stationary portal. Thus, time travel into the past is possible by carefully maneuvering between the stationary and moving mouths, taking advantage of the temporal discrepancy between them. Then, by returning to the starting point through the throat linking the mouths, this time difference could be successfully exploited.

Specifically, a past-directed journey could be achieved with the Caltech device by use of a special round-trip circuit. The trick is first to situate prospective time travelers close to the stationary mouth of the wormhole. If these explorers then make an exterior (free space) voyage to the moving mouth of the device, they find that the lapsed clock time there (relative to some original fixed point of reference that links the clock times of the two mouths) is much less than the clock time at the fixed mouth. Then by returning to their original location, this time *through* the wormhole throat, they emerge from the fixed mouth during an earlier era. They have, in effect, catapulted themselves back in time.

By altering the relative speeds and accelerations of the two mouths any amount of temporal displacement could theoretically be achieved by this method. Hours, years, even centuries could be whittled away from historical or personal time merely by passing through the right sort of wormhole. All that would be needed, in practice, would be a means of constructing a wormhole, a way of accelerating one of the mouths and a method of propelling spaceships between the two portals. These, of course, are not insignificant hurdles: it may be centuries before time travel is perfected. However, the Caltech team's results seem, in theory, to yield a highly plausible method of journeying through time.

Let's see now how a time travel excursion of the future might work. Imagine an astronaut starting out at a point near the static mouth of a wormhole time machine, when the year there at the mouth is 2500. She then begins a rapid acceleration of her ship, journeys through the intervening space, and ends up traveling close to the moving mouth at the same brisk speed. She notices that, since the clocks attached to the moving mouth are much slower, the year there is only 2490. Finally, by steering her craft into the second portal, she emerges through the first, back at her original location but during the year 2491 (assuming that traversing the wormhole takes one year). By a simple maneuver, she has effectively journeyed nine years into her own past.

As we can see, wormhole time travel is, at least in theory, straightforward, flexible, and safe. Some theorists, however, have ruled out the use of such time displacement devices on the basis of various philosophical dilemmas. In the next chapter we will discuss the fascinating debate that is currently taking place around this question.

CHAPTER 8

THE GREAT COSMIC BILLIARDS GAME

Cause and Effect

In politics, cause and effect are sometimes difficult to determine. During the frosty decades of Soviet-American rivalry that followed the Second World War and continued until recently, participants in scientific collaborations between rival blocs did their best to bore openings through the formidable barriers that divided their countries. Some might argue that these apertures hastened the ultimate crumbling of the wall between the superpowers and that scientific and other sorts of cultural exchanges among individuals of these nations paved the way for the consequent political breakthroughs. Others might contend that statesmen took the lead and that the negotiations carried out by senior politicians produced the treaties necessary for the thaw to occur. According to this view, scientists simply followed the path chiseled by the diplomats. Unfortunately, politics is not cut and dried; it's exceedingly hard to assign credit or blame for any sort of historical twist or turn.

It's precisely for this reason that physicists generally try to avoid enmeshing themselves in politics unless forced to do so. Physical scientists *thrive* on the absolute, on the nononsense, crystal-clear realm of universal facts, principles, and equations. In physics, the law of cause and effect is seemingly ironclad; every action produces a discernible reaction—eminently measurable and always occurring sequentially afterward. Certainly for the Russian, Igor Dmitrievich Novikov, and the American, Kip Stephen Thorne, the sturdy stepping stones of astrophysical theory were far more attractive than the slippery mud fields of political operations. The links fashioned by Novikov, Thorne, and their respective colleagues joined worlds perhaps as dissimilar as those that may someday be connected by interstellar wormholes. Their transcontinental collaboration linked universes as diverse as Pasadena, U.S.A., and Moscow, U.S.S.R. It is a tribute to the universal nature of astrophysical (and other scientific) inquiry that it can span such a formidable gap.

Of course, the recent revolutionary turn of events in Europe moved East and West far closer together and made it much easier for the Caltech and Moscow scientists to work with one another. During the tumultuous period of the past few years, cause and effect sometimes seemed to break down. But these scientists, more fascinated by physics than politics, were concerned about a much different sort of breakdown in the law of cause and effect.

The puzzling dilemma which brought together these theoretical physicists concerned the very nature of physical reality itself, specifically with regard to the implications of time travel via wormholes. The wormhole time machine model developed by Thorne and his fellow Caltech researchers seemed to rest on very solid theoretical foundations. Yet profound problems were found with this proposed device. The problems lay not with the equations or the calculations, but with the conclusions drawn by their work, namely that under certain circumstances causality could be violated by use of such a machine. Traditionally, physicists have rejected all theories in which effect can precede cause—and the Caltech model seemed to contain such a troublesome possibility. The logic seemed to strike a lethal blow to the entire theory of traversable wormholes: causality must not be violated; wormholes allow time travel; time travel allows causality violations; ergo, wormholes are impossible.

But the Caltech and Moscow physicists were not willing to give up so soon, certainly not at a time in which all around them conventional wisdom was being turned on its head.

No, in the era of the unlikely and the unimaginable, the Russian and American thinkers instead decided to put their minds together and invent the unprecedented: a cosmological theory *with* realizable time travel and *without* contradictions.

Closed Timelike Curves

In the past few decades, there has been a profusion of cosmological time machine models. What unites these models is the presence of an element, called a *closed timelike curve*, so baffling that physicists don't know quite what to make of it. All of these schemes, without exception, rely on this highly controversial technique in order for their device to work. Yet no one is quite sure whether this method is actually permitted by the laws of physics.

It is as if all new travel guides appearing in 1993 publicized a wonderful new scenic shortcut from New York to Washington, D.C. The only problem with this shortcut is that it required an illegal U-turn right in the middle of the New Jersey Turnpike! Suppose, then, that this route subsequently became extremely popular. The New Jersey government would be faced with one of two choices: either to forbid the shortcut or to legalize the U-turn.

Closed timelike curves are the cosmological equivalent of such a U-turn. Cosmologists today are faced with the disturbing choice of discarding these extremely promising theoretical models and keeping a long-held taboo against CTCs or else dropping the prohibition and, under certain special circumstances, permitting the models.

Traditionally physicists have held that CTCs are forbidden because they violate causality. Therefore, traditionalists argue, backward time travel exploiting the use of CTCs is simply a sign that the particular cosmological model involved is fundamentally flawed.

Other physicists, however, including Thorne, Novikov, and their colleagues, have argued that the appearance of CTCs in their models leads one to a far different conclusion: that some way must be found of successfully incorporating these anomalies into physical theory. They regard CTCs as problematic, but ultimately workable, components of cosmology.

What is it about closed timelike curves that causes so much trouble? As Nick Herbert points out in *Faster than Light: Superluminal Loopholes in Physics*, CTCs represent "loopholes" in relativity: ways in which the prohibition against faster-than-light travel can be circumvented. Consequently, they permit intrepid explorers ways of avoiding the normal chain of cause and effect embodied in ordinary subluminal interactions. Many physicists find the idea of the existence of such escape routes to be philosophically troubling and reserve judgment about CTCs until they are experimentally detected or definitely proved to exist in some other way.

To understand CTCs, as well as what makes them so disturbing, one must first examine what special relativity has to say about limits to communication between distinct entities. These limits differ, depending on the region of the universe under consideration.

In normal, flat (Minkowski) space-time, away from massive objects such as black holes, interactions between particular events take place within prescribed frameworks, called *light cones*. These abstract structures are a direct consequence of the special relativistic prohibition against exceeding the speed of light.

Light cones can be best pictured by imagining a three-dimensional space-time diagram with time as the vertical axis (up, out of the page) and two of the spatial dimensions as the horizontal axes (along the page, perpendicular to each other). To envision this, picture a flat infinite chessboard, placed on a table. This chessboard represents *space*, with the two dimensions of space depicted by the two directions along the board (space actually has three dimensions, but we shall ignore the third for the purposes of this analogy). Then, the up and down direction, above and below the chessboard, represents time.

Now imagine that each object in the universe is a king on the chessboard. A chess king can only move one square at a time, albeit in any direction. Let's call this velocity "the speed of chess." Thus, no object in the chess universe can exceed the speed of chess, just as no body in the real world can travel faster than light speed.

Suppose we wish to indicate on our diagram the farthest extent to which a king may travel for any given number of moves (that is, length of time). In zero moves, the king may not move at all. We indicate this by a point centered on the position of the king. In one move, the king may go to any of its neighboring squares: forward, backward, sideways, or diagonally one square. We can represent the full extent of one-move behavior by circling this set of possible moves. More vividly, we place a ring of one-square radius an inch above the chessboard. We do the same now for the two-move possibilities. We ask, How far may the king travel in two moves? Clearly, this can be indicated by a ring of two-square radius. We therefore place a ring of two-square radius an inch above the ring of one-square radius. Extrapolating, the maximum extent to which a king may travel after three, four, five, or any number of moves can be represented by rings of three, four, five, and so on, squares each. If we place each ring the number of inches above the board that corresponds to the number of moves played (the time, so to speak), we find that the rings form a circular cone, centered on the king, projected perpendicularly above the board. This is the *light cone* for the chess game; it represents the maximum extent of the king's travels for any particular time.

Viewing the light cone gives us a great deal of information about the game. Specifically, it indicates the domain of motion allowed for the king. Inside the cone is the region in which the king might move during a given match: the realm of possibility over time. It's clear why, technically speaking, motion within the light cone is called timelike. The cone itself is the zone of maximal motion, usually referred to as the *null region*—neither timelike or spacelike. Outside the cone is the forbidden region, called the spacelike domain, for which no possible moves, over time, can exist. Thus the light cone serves to divide space-time into three regions: timelike, spacelike, and null.

Moreover, this space-time diagram offers additional information about the game. The boundaries of the light cone also depict whether or not *interactions* may take place. A king is allowed to capture any piece with which it comes into contact. Therefore, the maximum extent of the king's motion per move, indicated by the light cone, is also the maximum extent of its interaction. For example, a piece located three squares away from the king may be captured by it in no fewer than three moves—a fact clearly depicted by the cone.

This is not the whole picture of space-time, however. The forward-time light cone, pivoted on the chessboard and rising up from it, is only half of the story. To complete the diagram, the cone must also be extended down below the board. Hence, the bottom half of the representation of the king's maximal motion is an inverted cone attached below the original one.

Like the upright section, the upside-down portion of the light cone also conveys a considerable amount of information about the king's range of achievable behavior. In the latter case, the information depicted is that of possible past motion of the piece. For a given number of moves in the past, the king could have emerged from any square within the cone's interior region (the timelike past) or, at the very most, from the zone indicated by the cone itself (the null past). However, it's clearly impossible for the piece to have emerged from squares outside the cone's domain (the spacelike past).

Using this picture, we might also view the possible interactions between two chesspieces. First we extend the model to assume that each piece has its own light cone, containing both future- and past-directed segments, corresponding to its complete range of behavior. Then we examine the configuration of the cones to determine possible connections between the players.

Now suppose that we wish to find out, using this diagram, whether or not two pieces may interact (one presumably capturing the other). The possibilities are clearly indicated by the cones. First, if the cones don't interact during a particular interval of time, then interaction between the pieces is impossible for that period. Second, if the future parts of two cones intersect, then a capture will take place some time in the future. Likewise, if the past sections of the cones touch, then an interaction has already occurred in the past. Finally, if the future section of the light cone of one piece crosses the past section of the other, then an interaction occurs in the future of the first piece but in the past of the second.

This chess game analogy is highly appropriate when considering the behavior of objects in flat space-times. There, special relativity, with its faster-than-light prohibition, exactly models the motion of particles on a large scale. Of course, rather than speeds of "one square per move," light cones represent maximum velocities of 186,000 miles per second (mps). For flat space-times, these cones are diagrammatically depicted as being exactly vertical, that is, centered on the time axis.

Close to massive celestial bodies, however, special relativity must be supplemented by the dictates of Einstein's theory of general relativity. As we have discussed, this theory asserts that massive objects, such as stars, quasars, and black holes, curve space-time in their vicinity. Thus space-time in such a region may best be modeled by a curved surface, such as the exterior of a saddle or a beachball, rather than by a flat plane, such as the top of a chessboard. (Once again, note that space-time is really *four*-dimensional, like, for instance, the surface and exterior of a *four*-dimensional beachball. However, since virtually no one can picture such an object, let's stick for now to our three-dimensional simplification.)

Even in a region of curved space-time, light cones must be considered as perpendicular to the surface. Thus, under such circumstances, all of the cones in the vicinity are tilted away or toward each other, depending on the surface curvature. For example, if the space-time of a region looks like the exterior of a beachball, then the light cones can best be depicted as spikes protruding from the surface of the ball.

This tilting causes the nature of interactions between nearby objects to alter considerably. Inside of behaving (as in the case of flat space-time) like bristles on a hairbrush— upright, evenly spaced, and parallel—the light cones act more like swaying stalks of wheat—leaning this way and that. This indicates that a far more complex set of causal relationships is possible between events. It may be the case, for instance, that in more cases future parts of some light cones have become connected to past segments of other cones, indicating that some objects in their future now affect others in their past.

Another way that this tilting effect might be produced is by rotating or otherwise accelerating the space-time. According to general relativity the effects of acceleration mimic those of gravity and any phenomenon produced by one may be reproduced under the right circumstances by the other. For example, it is impossible for passengers in completely enclosed spaceships to determine whether they are being accelerated by a rocket engine or propelled at the same rate by gravitational attraction. So, for instance, all of the sun's gravitational effects may be reproduced by accelerating the space-time nearby it in the appropriate manner.

Among the cosmological and astrophysical models we have already discussed there are many examples in which light cone structures are distorted, either by gravitational attraction or by rotational motion. In each of these cases, this distortion leads to predictable changes in the causal relationships among events occurring on neighboring points. In other words, objects located in these anomalous regions interact differently than they would under the conditions of flat space-time.

Consider, for instance, Gödel's rotating universe model. Recall that by traveling in a circular motion around the axis of such a universe astronauts can travel backward in time. The reason for this is the special configuration of light cones in such a space. The rotation of Gödel's universe causes all of the cones in the vicinity to tilt in the direction of spin. As in the case of toppling dominoes, for sufficiently high rotation, the cones begin to "lean" on each other and become more and more spacelike (horizontal, according to the space-time diagram). Consequently, for sufficiently large spin velocity and radius from the axis, the cones become aligned such that the future section of the light cone of one location intersects the past section of the light cone of the adjacent location. Thus, interactions take place between the future of each point and the past of its neighbor. In other words, communication and travel take place backward in time as one moves or sends a signal cyclically from one point in the universe to another.

Eventually, if one were to travel in a complete circle in a Gödelian universe, one would reach the starting point of one's travels. However, because of the continuous backward motion in time, from the future cone of one point to the past of the next, one would reach the initial location some time in the past. The light cone structure of Gödel's universe can thus be best described as something like the shutter of a camera, with the past and future parts of the light cones layered on top of each other in a circular pattern. This special pattern enables signals or objects to travel around the circle, moving all the while from future to past. In this manner, backward-in-time messages can be sent around the ring without violating the Einsteinian prohibition of faster-than-light transportation or communication.

The peculiar light cone configuration displayed by Gödel's model is an excellent example of a closed timelike loop. The CTC here is the path taken by travelers or light signals as they orbit around the axis of the universe at a sufficiently large radius. It is clear that such an orbit involves achievable backward time travel, accomplished by careful threading of joined future and past segments of light cones.

In a similar manner, CTC paths can be found in the vicinity of other cosmological time travel devices such as the Caltech wormhole model and Gott's cosmic string approach. In each of these examples a way can be found for an explorer to travel back to his or her past by using tilted light cones.

Closed timelike curves offer the enticing promise of time travel. However, this promise is tempered by the lack of proof of their actual physical existence. A general, comprehensive solution of Einstein's equations of general relativity, revealing all of its quirks and anomalies, has yet to be found. Until it is, only two avenues are open to theoreticians: rough, incomplete mathematical approximations and detailed computer simulations.

Generally, physicists prefer to solve deep questions, such as the existence of CTCs, by use of pencil and paper, rather than computers. There is a long tradition in physics of finding exact solutions to problems in the form of algebraic formulas. These results yield information normally absent from computer data sheets. The equation $E = mc^2$, for instance, is far more significant than a computer output sheet simply containing energies and masses of particles.

On the other hand, in the case of wormholes, black holes, and other such objects, analytic (not computer-generated) solutions require an enormous amount of oversimplification. For example, in all of the traversable wormhole models that we have considered we have made several simplifying assumptions, including the requirement that they have either spherical symmetry or another fairly basic sort of geometry. These assumptions were necessary to reduce the difficulty of solving Einstein's equations of general relativity for these models. In a more realistic portrait of these objects, all possible types of geometries would need to be considered—a feat simply not possible by hand calculation.

Currently, a number of scientific researchers are busy attacking the problem of whether or not CTCs actually exist by use of massive supercomputers and extremely powerful software programs. The majority of these projects are specifically examining the region surrounding spinning (Kerr) black holes to look for examples of causality violation. Computational work has yet to be done on the question of CTCs and generalized traversable wormholes. So far, scientists have found no conclusive evidence that CTCs exist in nature. However, it is hoped that continuing computer studies will soon solve the question, one way or another.

If they do exist, closed timelike curves represent a radically new sort of phenomenon in physics: they are the first examples of the interaction of the future of some region in space with its past. Consequently, when CTCs are exploited for use in cosmological time machine devices, such as the ones developed by Morris, Thorne, and Yurtsever, these curves embody a supreme violation of the law of cause and effect.

It is for this reason that the Cambridge physicist Brandon Carter calls these pathways into the past "vicious circles." For him, CTCs exemplify mathematical nuisances in applied general relativity, ones that should be eliminated, if possible, by careful reinterpretation of the theoretical results. For example, one can simply assume that travel through wormholes (or black holes) never returns you to any part of your own universe. In that case, an astronaut journeying through such a device couldn't interfere with his own past because return to his starting point would be impossible. Thus, it is clear that CTCs can be eliminated if one takes the radical step of completely discarding the possibility of round-trip wormhole voyages.

This argument doesn't work for Gödel's universe, however. Round-trip voyages are clearly possible in that case by traveling in a circular loop around the axis and hence forming a complete CTC. Carter calls space-time regions in which CTCs cannot be eliminated by simple reinterpretation of results "flagrantly vicious sets." He contrasts these with areas, called "trivially vicious sets," in which a simple ban on round-trips completely eliminates the possibility of closed timelike loops. The latter includes the vicinities of spinning black holes and traversable wormholes.

It seems, however, rather arbitrary to make such a distinction between seemingly achievable and nonachievable CTCs. There is no reason whatsoever to assume that passengers traveling through wormholes would never be able to get back to their starting points. As long as even a remote possibility of cyclical wormhole journeys exists, it is clear that traversable wormholes are perfect conduits of closed timelike curves, threaded, in part, through their throats.

In the absence of evidence to the contrary, we shall make the perfectly reasonable assumptions that the wormhole time machine models developed by Morris, Thorne, and Yurtsever are achievable in nature, at least in theory, and that the closed timelike curves found in these models are valid outcomes of general relativity. It is hoped that a thorough experimental test of these suppositions will someday be possible; until then only careful logic should limit our imaginations. With this premise in mind, let's now look at the direct philosophical implications of the existence of wormhole time travel and closed timelike curves.

Time Travel Paradoxes

The creation of a wormhole-based time machine would be a most extraordinary event that would change history forever. Scientists could use the unit for fact-finding trips into the future, while historians could use the device for revealing reexaminations of the past. Yet it has long been known that the existence of time travel leads to perplexing logical puzzles and paradoxes.

Not all time travel would involve philosophical paradoxes. Journeys into the future of the Wellsian sort would contain no contradictions. In Wells's *The Time Machine* the protagonist sets out into the future, stays for a brief interval, then returns to the present. Although his presence certainly alters the events of the world to come, such a change violates no laws of logic or physics, since the order of cause and effect isn't interfered with in the story. Causality can only be violated when signals are sent from the future to the past, not vice versa.

On the other hand, in Ray Bradbury's science fiction opus "A Sound of Thunder" causality is tampered with in a disastrous manner. The story begins right after a pivotal presidential election in which a fascistic candidate is barely defeated. Relieved by this fortunate turn of events, the tale's protagonist, named Eckels, decides to take a holiday. He finds out about a new "time safari" adventure tour, in which vacationers are whisked away into the prehistoric past, given rifles, and sent out to hunt dinosaurs.

Heading out on the tour, Eckels receives careful instruction about the dangers of time travel. He is warned to shoot only the carefully marked animals, to stick closely to the designated trail, and to avoid contact with plants and insects. Otherwise, because of interference between present and past, history could be rewritten. Only the dinosaurs that are expected to die soon anyway are permitted to be killed, and they are marked with red paint. By shooting only these specially labeled creatures, it is presumed that no changes in history could occur.

Unfortunately, Eckels makes a horrible blunder. Frightened by the approach of a tyrannosaurus rex, he unintentionally strays off the path. Tramping through the mud, he accidentally kills a small butterfly. Not immediately realizing this, he returns to his present, via his time machine, only to find that the world has been radically transformed beyond belief. The death of the butterfly has triggered a chain of events that have been magnified over time, leading to an enormous difference in history. This discrepancy becomes apparent to the traveler when he discovers, first, that the English language has been altered and, second, that the presidential election outcome has been decided, not in favor of the democrat, but instead in favor of the authoritarian. Eckels realizes, to his dismay, that his contact with the past had been much riskier, and ultimately far more destructive, than he had expected. Finally, Eckels is shot—punished for tampering with history.

In Philip K. Dick's classic science fiction tale *The Man in the High Castle* a similar theme is developed in which the past is altered by an exterior force. The novel is set in an alternative America, where Japanese and German forces reign supreme. A massive disruption of the stream of time has resulted in a bizarre turn of events in which the Allies have lost the Second World War and the Axis powers have established global hegemony. As in Bradbury's frightening story, it's made all too clear how a relatively small change in history could have long-lasting global consequences of the most ghastly sort.

The idea of a parallel version of earth in which Hitler has conquered the planet is quite a common one. A large number of short stories and novels on this theme have recently been collected in Gregory Benford and Martin Greenberg's *Hitler Victorious*. Apparently, the notion that the outcome of the Second World War could have been quite different resonates with frightening clarity.

This topic is echoed in Ward Moore's disturbing novel *Bring the Jubilee*, similarly set in a parallel version of America but concerning a different historical period. In this case, a time traveling scholar arrives back in the year 1865 and ultimately changes the course of the Battle of Gettysburg. This results in an alteration of the outcome of the Civil War and in a subsequent branching of history. The ironic twist to the story is that the scholar is from a world in which the Confederacy has been victorious over the Union but ultimately, by his inadvertent tampering, creates the conditions for our present society, in which, of course, the South lost the Civil War.

If, by use of devices such as Morris, Thorne, and Yurtsever's wormhole model, time voyages become possible, perhaps even commonplace, what is to prevent foolhardy wanderers from tampering with history in an ultimately destructive manner? What is to stop mad Fascists from returning to the early 1940s and providing vital information for a military victory by the Nazis or to deter disgruntled historians from supplying Genghis Khan with a hydrogen bomb? And if, as in Bradbury's story, accidentally crushing a mere insect would cause historical events to topple, one domino after another, how could anyone predict what would happen if time machines entered into general use? Clearly, some means of damage control would have to be developed; otherwise, catastrophic changes would wreak havoc on the world.

Perhaps history somehow monitors itself and corrects all attempts at radical alteration. It could very well be that any attempt to tamper with time would be met with frustration as intended historical changes were negated by turbulent counterforces. For example, if Hitler had been assassinated, there is a chance that another leader might have taken his place and initiated a similar, but not identical, series of events. World War II might still have been started by the alternate leader, albeit under a different set of circumstances and a different timetable. The United Nations might still have been formed, Truman might still have become president, Stalin might still have died when he did. Consequently, the year 1990 might have, in fact, been almost exactly the same as it was. Hence, it is conceivable (but clearly unlikely) that the results of even such a radical change as this might still have become muted over time as a result of random historical countercurrents.

The idea that history might be self-corrective is one of the prevailing themes in Fritz Leiber's "Change War" chronicles.

In his stories, two rival, time-traveling factions, the "Snakes" and "Spiders," constantly attempt to change history to the other group's disadvantage. However, members of these groups eventually discover that tampering with reality is ultimately futile because, according to "The Law of Conservation of Reality," all changes must be negated over time.

Clearly the domino effect and the conservation of reality principle are two extremes of the same spectrum. Most likely, the results of any interference with time would vary, depending upon the circumstances. In a few cases, stepping on an insect might lead to a toppling of governments. However, in most instances, nothing of the sort would occur. On the other hand, the assassination of a political leader would almost certainly lead to long-lasting repercussions. Only lucky (or unlucky) fate would prevent events such as the boyhood assassination of Lenin from radically altering the political history of the twentieth century, it would seem. Thus, wormhole time travelers would never know the true effects of their interference until it was too late.

We've looked at some of the more unpleasant or unsettling aspects of time travel. Yet we still haven't touched on the key philosophical concerns that compel many scientists to abandon hope for realizable trips into the past. These latter theoretical issues, grouped under the general heading "time travel paradox," are the far more difficult ones with which to deal.

A time travel paradox is not just an interference between the future and the past, such as if someone were to go back to the 1910s and kill the young Hitler. That sort of event would be relatively simple to deal with logically; one might simply suppose that the universe would start to undergo a different sort of evolution—one without Hitler or the Nazi regime. As long as the time traveler himself wasn't affected by the events that transpired as a result of this change, there wouldn't be any contradiction whatsoever between alternate versions of reality: the first version of truth would simply disappear.

Imagine, however, the following truly paradoxical scenario: It's the year 2500. Unfortunately, one of the new wormhole time machine devices in the outer solar system has just been captured by Benito Herring, a prominent neo-Fascist with an insane hatred of all people from Andorra. Herring, in a fit of rage, has decided to go back in time and destroy all Andorrans. He travels back to the year 2000 and unfortunately succeeds in eradicating the mountain kingdom.

Herring, however, is strangely unaware of one important detail of his ancestry: he's 100 percent Andorran. By obliterating Andorra, he has in effect murderd all of his ancestors. Therefore, if there is any sort of cosmic justice, he should rightfully disappear from the earth, a victim of his own stupidity. How could anyone exist without any of his ancestors?

But if Herring popped out of existence like an expired candle, then who would carry on his horrific program? Clearly, he wouldn't be around in 2500 to take over the wormhole and trek into the past. Without him to initiate such a plan, certainly no one would set out to accomplish his uniquely fiendish goals.

Therefore, in that case, the people of Andorra might breathe a massive sigh of relief (if they only knew what fate might have had in store for them). They'd go on living their lives exactly as they would have—marry, have kids, and thus perpetuate their people—until one summer day in 2470, a tiny tyrant by the name of Benito Herring is born to their descendants and grows up remarkably unaware of his Andorran heritage, and, furthermore, feeling a barbaric hatred of all Andorrans.

One day, Herring decides to travel back in time and launch a heinous scheme regarding a certain group of mountain people.

And so on, and so on.

The point here is that this story could keep going on forever as a continuous fluctuation between two logically impossible outcomes: in the first case, the Andorran people are destroyed, but Herring is born anyway; in the second case, the Andorran people survive, yet Herring is never born. Clearly, there is a logical flaw here, a paradox built into the very nature of time travel. The stream of history, which we usually expect to be straight and continuous, seems strangely twisted in on itself.

Time travel paradoxes are a common source of perplexity in science fiction. Often, such contradictions are barely avoided—they are resolved just in the nick of time. For example, the recent *Terminator* film series involves a society of the future that is dominated by computers and robots. When a band of rebellious humans, led by a man named John, attacks the machine-based hierarchy, the computers order that "terminators," robots programmed for murder, be sent back into the past, first to try to kill John's mother before John is born, and second (in the sequel) to assassinate John while he is still a boy. Fortunately, both attempts fail and the rebellion takes place.

Critics of the films have pointed out, however, an obvious flaw in the series premise. Suppose the terminators were to succeed in killing John while he was still a young boy. Then, of course, there would never be a rebellion in the future against the computer rulers. Therefore, they wouldn't have to send terminators back into the past and John wouldn't, in fact, be assassinated. Thus, we'd have another example of an unresolvable time travel paradox.

A second type of dilemma involves sending objects a short interval back in time. If these objects have been temporally transported, why haven't they been seen already? For example, suppose you decide to use a time machine wormhole to propel a newly manufactured spacecraft one year back in time. In that case, why hasn't that ship been observed a year before?

Or consider the following situation: Herbert George, an extraordinarily lazy man who happens to own a time machine, decides one morning that he wants breakfast in bed. "No problem," he thinks. "Tomorrow I'll make breakfast and send it back in time to the present. I'll program the controls of the machine to have the breakfast appear this instant on my night table."

Sure enough, Herbert looks to his side and notices a steaming hot breakfast of bacon and eggs on his night table. The mere decision to send the food back in time has provided him the means of sating his hunger without having to move a muscle.

In this example there is an obvious dilemma: what causes

the action of putting breakfast on Herbert's table? Is it Herbert's intentions, his thoughts, by themselves? If so, doesn't this imply that, given the existence of time travel, our minds could work magical feats?

This theme has been exploited many times by short story writers and Hollywood producers. The most recent fictional incident that comes to mind takes place in the farcical time travel comedy *Bill and Ted's Bogus Journey* (the sequel to *Excellent Adventure*), in which the two heroes, after deciding that they need certain tools to net a villain, agree that they will eventually pick up the supplies and go back in time with them. Sure enough, the instruments appear out of nowhere.

The problem with such a "labor-saving" technique is that there are no guarantees that the intended time voyage would ever come to fruition. If the journey never takes place, then a paradoxical situation is created. Suppose, for instance, that lazy Herbert George never makes his breakfast and never uses his machine to transport it. From where, then, did the breakfast come?

This strange paradox is the focus of Fredric Brown's short story "Experiment." In this tale, Professor Johnson, creator of the world's first time machine, holds a brass cube in his hand. At exactly three, he tells his colleagues, he will place the cube on his time machine's platform and send it five minutes into the past. Sure enough, at five minutes to three, the cube vanishes from the professor's hand and appears on the platform, having been sent back in time.

"But," says a colleague, "what if, now that it has already appeared, you should change your mind and not place it there at three o'clock? Wouldn't there be a paradox of some sort involved?"

Professor Johnson finds this idea to be interesting and doesn't put the cube on the platform at three. What happens? The cube remains, but the rest of the universe, including Professor Johnson and his colleagues, disappears.

These science fiction stories and thought experiments show us some of the serious problems involved in schemes for backward time travel. But what does this have to do with real science? Nothing—that is, if time machines remain simply fictional plot devices. Everything, however, if one is considering the merits of realizable traversable wormhole schemes that may someday be used for space or time travel.

Morris, Thorne, Yurtsever, Novikov, and their wormhole-constructing colleagues take these paradoxes extremely seriously. For them, these are not just plot twists of inventive science fiction stories. The whole question of wormhole viability rests on whether the existence of wormholes is consistent with the laws of physics and the constraints of logic. Otherwise, there is sufficient reason to abandon all hope of ever building these gateways.

Could the use of a wormhole lead to contradictions in physics, they wonder? They consider, in one of their journal articles, the example of Schrödinger's cat. Recall that this is the thought experiment in which a cat is placed in a box and is considered to be in a mixed state of aliveness and death until the box is opened and its "wave function" is said to "collapse" to one of the two possibilities (see Chapter 5). Suppose an advanced being measures and determines Schrödinger's cat to be alive at a certain time, say noon, thereby collapsing its wave function into a live state. After this observation, he goes backward in time, via a wormhole, and kills the cat before noon, collapsing its wave function into a dead state. When the clock strikes twelve, would the cat be alive or dead (and what would be the state of its wave function)? Clearly, this is another superb example of a time travel paradox.

The Principle of Self-Consistency

It is because of paradoxes and dilemmas such as these that the Caltech and Moscow physicists have decided to use a new approach to address these problems. They've developed a means of ensuring that all wormholes that they design obey the basic rules of logic and do not allow for unresolvable inconsistencies.

The new litmus test that these scientists use is called the *principle of self-consistency*. This is a way of allowing closed timelike curves but excluding the possibility of physical discrepancies. They urge that this principle be incorporated into physics, much like the law of conservation of mass and energy or the law of the constancy of the speed of light in a vacuum. Only the introduction of such a constraint would permit CTCs, and hence wormholes, to exist. As Novikov, Thorne, Morris, and Yurtsever state this in one of their most recent papers (also authored by John Friedman, Fernando Echeverria, and Gunnar Klinkhammer):

> Can CTCs be incorporated into the laws of physics without producing unacceptable causality violations? The answer to this question depends, of course, on one's definition of "unacceptable."
>
> The only type of causality violation that the authors would find unacceptable is that embodied in the science fiction concept of going backward in time and killing one's younger self ("changing the past"). . . .
>
> We shall embody this viewpoint in a *principle of self-consistency*, which states that *the only solutions to the laws of physics that can occur locally in the real universe are those which are globally self-consistent.*

In other words, if some action or motion occurs locally—that is, over a short distance or short period of time—it had better be consistent with the overall evolution of the whole universe. If one writes out equations that determine the behavior of the entire cosmos from beginning to end, then all CTCs must somehow tie in to the solutions of these equations.

Let's consider an analogy to make this point clearer. Imagine the beginning of time to be a train station and the end of time to be a second station. A single straight track, connecting the two terminals, represents the continuous evolution of the universe over time. Note that there is no "backtracking"—indicating that the behavior of the cosmos is unidirectional, that is, one-way over time.

Unfortunately, the railway is operating at a loss and the town planners decide to rip up the track and build a single-loop roller coaster in its place. They want to keep the same two stations and build the coaster between them. Engineers go to work developing a set of plans for the new attraction.

To continue our analogy: here, the plans represent the equations dictating the evolution of the universe. The loop of the roller coaster stands for a CTC—a curving back in time. The fact that the coaster must begin at one station and end at the other symbolizes the problem that physicists must face: to develop models of the universe that evolve from a set of initial conditions to a set of final conditions.

The engineers produce a number of schemes for the roller coaster and must decide which one to use. To determine which blueprints are valid and which are defective they decide to use the litmus test developed by the engineering firm of Morris, Thorne, and Novikov, Inc., called the "principle of self-consistent coasters." This rule states that looped coasters are not allowed if they cause the track to deviate from the initial or final stations. Only coasters which precisely link both terminals are valid. Therefore, the designers decide to use only those loops that orient the track in the correct directions: toward both train stations.

To apply the principle of self-consistency to physics we ask the following question about each scheme for time travel: Do the CTC loops produced by the model throw the universe off track, so to speak, or do they steer the universe in the same general direction that it would have taken anyway? That is, do they fit in with an overall plan for all of time?

Thus, in the case of traversable wormholes, we insist that all models developed must fit in with a reasonable outline of the development of the cosmos as a whole. If a wormhole scheme fails to correspond to this global portrait of reality, then we may safely reject it.

Under these guidelines, backward time travel is allowable if it doesn't alter the natural course of the world. The only changes to the past that are allowed are those that were meant to be. As Ian Redmount of the British magazine *New Scientist* puts this:

> The evolution of a physical system should be self-consistent, even when you include influences from the future acting back in time. This means that if you travel back in time and attempt to shoot your parents before

your birth, your gun misfires or you miss; the sequence
of events already includes the effects of your attempt.

Although the principle of self-consistency has never been
applied to theoretical physics, it has certainly played an
important role in science fiction stories as a way of avoiding
paradox. For example, in Michael Moorcock's *Behold the
Man* an emotionally disturbed ultrareligious individual ob-
tains access to a time machine. Journeying back to the time
of Jesus, he is startled to find no trace of the Saviour. After
his time machine is destroyed and he is stranded, he inadvertently begins, step-by-step, to reproduce all of the events
recorded in the Gospel, with his playing the role of Jesus. It
turns out, of course, that he is the real Jesus and that Biblical
history has been preserved. Thus, according to this tale, the
history of the universe has remained self-consistent.

Another classic example of self-consistency in science fiction is Robert Heinlein's amusing tale "All You Zombies."
In this story there are no fewer than three CTC-type loops,
and yet the world portrayed is entirely compatible with itself
(paradox-free). The plot goes something like this (it's hard
to do justice to the numerous twists and turns): An orphaned
girl, who has just reached childbearing age, is seduced by
a strange man and has a child by him. Mysteriously, the
man disappears, and the child is stolen from her nursery.
Furthermore, to make matters worse, during the delivery a
mysterious complication mandates that she undergo a sex-
change operation. Hence, she becomes a man, albeit a
peculiar-looking one. Some years later, she/he meets a time
traveler. He whisks the unusual man back into the past and
incites him to molest his earlier self. Then he steals the child
and places it back in an orphanage. It thus turns out that the
time traveler, orphaned girl, and seducer are all the same
person. In other words, the girl is her own mother, father,
and abductor in an entirely self-consistent loop!

Examples of self-consistent circuits in films abound. One
entire film series, the *Planet of the Apes* movies, forms a
complete and consistent closed timelike loop. In the first
film, three astronauts travel through time to future earth and
find it to be dominated by intelligent apes, who rule over

devolved humans. The next film in the series involves the conflicts between apes and humans. In the third film, two of the apes, Zira and Cornelius, find the astronauts' spaceship and use it to travel back in time to present-day earth. There, Zira gives birth to Caesar, who naturally turns out to be intelligent, like his parents. Then, in the fourth part of the series, Caesar grows up and inspires all present-day apes to rebel against humans and to take over. Finally, in the fifth and last *Planet of the Apes* movie, the apes form a society that clearly rules over the humans. Hence the entire series of films forms a complete and inseparable epic; no part of it could be left out without the series's being somehow flawed. And, because of the careful regard for self-consistency, no paradox exists in it whatsoever.

Even when a paradox is successfully averted, the results can be unsettling. Consider, for example, "Find the Sculptor," a story by Sam Mines in which a scientist creates a time machine, travels five hundred years into the future, and finds a statue there of himself, erected in honor of the first time traveler. He then uproots the monument and takes it back to his own time as proof of his successful journey. Consequently, the statue is set up to commemorate his voyage. The question Mines asks at the end of the tale is, When was the statue made? Clearly, though this story is self-consistent, it is troubling.

*W*ormhole Billiards

As the statue story indicates, the principle of self-consistency doesn't eliminate all troubles involving wormhole voyages into the past. But, as the traveling salesman in the musical *The Music Man* points out, there are troubles everywhere. For River City, the idyllic town depicted in that play, trouble appears in the form of pool. As the salesman puts it: "You've got Trouble . . . With a capital T that rhymes with P that stands for Pool."

For Kip Thorne and his co-workers, however, trouble comes in the form of *billiards*. One might wonder what billiards, of all things, has to do with wormholes. Then again, what do cats have to do with quantum mechanics?

Thorne asks us to imagine an enormous billiards game involving a gigantic billiard ball and a traversable cosmic wormhole with two closely spaced mouths. The wormhole is of the time machine sort, and it propels all objects back into their past. Hence, when the ball is hit into the first mouth, it reemerges through the second mouth back in time.

Now suppose that the giant ball is sent into mouth 1, reappears at an earlier time out of mouth 2, and then collides with itself. As a result it knocks itself away from its initial route and it never enters the wormhole. Thus, the paradoxical situation that it can never bump into itself occurs.

There is a clear discrepancy here between two different versions of reality. The initial conditions of this problem, including the starting position and speed of the billiard ball and the starting location of the wormhole's mouths, are such that no single consistent solution is possible. On the contrary, there is one trajectory involving the ball's entering the first mouth and another involving the ball's not entering it. Yet the conditions are such that, paradoxically, if it *does* enter the mouth, it must collide with itself and *prevent* itself from entering it.

Stated in this manner, the billiard ball puzzle is a clear example of *self-inconsistency*. It represents the sort of dilemma that the principle of self-consistency is intended to weed out. Therefore, as Thorne points out, it is equivalent to the "killing your younger self" paradox and poses no additional problems, aside from restricting the range of initial conditions.

However, it is easy to imagine a perfectly self-consistent cosmic billiards situation that nonetheless involves another sort of logical dilemma. This additional problem was alluded to in our discussion of the story "Find the Sculptor": the requirement for consistency doesn't necessarily remove ambiguity.

It seems that in Sam Mines's tale there are *two* possible self-consistent universes. In one case, detailed in the story, the time traveler journeys to the future, discovers the statue of himself, takes it back in time to the present day, and then witnesses its being placed in the ground. Thus the statue travels around full circle in a closed timelike loop. The second self-consistent possibility is that the traveler journeys to the future, doesn't discover any statue, returns to the present, and no statue is erected to him. This is a perfectly reasonable, self-compatible, alternative version of the story, one in which no statue travels around a loop. As in the roller coaster analogy there are two possible links between beginning and end: one with no loop and the other with a single, well-designed loop. Well-designed here means obeying the principle of self-consistency: guaranteeing the unity, from start to finish, of the entire project.

Thorne and his fellow researchers have noticed the same sort of dichotomy in the billiards problem: given the right set of initial conditions one can have two self-consistent solutions of zero loops and one loop, respectively. Let's now see how the billiard ball puzzle might yield two possible results:

Imagine that the gigantic billiard ball is aimed directly between the two mouths of the wormhole and is set to follow a straight course. Naturally, without further impediment, it keeps on going in the same direction forever. This is the *zero loop solution* to the problem.

However, there is a second logical possibility. The ball could be knocked into the first mouth of the wormhole, travel back in time, and emerge from the second mouth. On leaving the wormhole it could be knocked into a straight path again. Thus forced into its original direction, it would keep on going forever. This is the *one loop solution*: note that it's identical to the zero loop solution except for the path twist in the middle, between the wormhole mouths.

What, then, knocks the ball into the first mouth of the wormhole? And what hits it back into its original trajectory? The answers to these two questions are related: in the first instance a version of the ball before it enters the wormhole (the "unprocessed version") is shoved into the first mouth by the version of itself that has already traveled back to the past through the wormhole (the "processed version"). In the second case, the version of the ball that has just emerged from the second mouth (the "processed version") bounces off the version of the ball that is about to enter the wormhole (the "unprocessed version") and is propelled away from the collision site along a straight path. Thus, both collisions and both participants are really one and the same.

Let's now consider a second equivalent example of dual self-consistent solutions to clarify this remarkable puzzle further. Instead of billiards, let's imagine an even more prosaic situation involving an ordinary man on his way to work. We assume, as in the earlier case, that time traveling wormholes exist.

Robert Preston, a salesman, lives and works in River City and takes the same route from his home to his office every day. He follows a perfectly straight route, passing, at one point, midway between a phone booth and a bush. This is the zero loop solution of the problem of mapping Robert's route from home to work.

What, then, is the one loop self-consistent solution to the task of plotting Robert's walk? The second possibility is the following:

Robert Preston, a salesman, lives and works in River City and takes the same route from his home to his office every day. He follows a perfectly straight route, passing, at one point, midway between a phone booth and a bush. One day, however, when he nears the bush, a hostile man, looking and sounding exactly like Robert, jumps out of the shrubbery and shoves him into the phone booth. The imposter then turns and keeps going along Robert's original route. Finally he reaches Robert's work, where he is naturally mistaken for his double.

Meanwhile Robert, quite frightened after being pushed into the phone booth, finds that he has been suddenly transported into the bushes (and one minute back in time as well, via an experimental phone company wormhole). He emerges from the shrubbery, then, suddenly noticing the imposter walking up (who is really Robert a minute earlier!), shoves him into the phone booth. Robert then turns and keeps going along his original route. Finally he reaches his work, where he decides to forget about the whole strange incident.

Notice once more that these are two alternative self-consistent solutions to the problem of determining the motion of an object, namely Robert. In both the billiards puzzle and the salesman incident there are two variant histories involving the same starting and finishing states.

Thorne and his co-workers find the existence of two self-consistent solutions to this type of puzzle to be most intriguing. Wormhole time travel, if realizable, appears to allow for multihistory approaches to trajectory problems, much as in the case of quantum mechanics.

Recall that in quantum physics, which is believed to be the most accurate representation of microscopic particle behavior, it is impossible to determine the position and momentum of an object at the same time (see Chapter 5). Hence, it is similarly impossible to map out the precise trajectory of any particle, given knowledge of its initial and final states. One can only assume that the object has a finite probability of having traveled along every possible path; these probabilities are indicated by the wave function of the particle.

It is as if a carnival barker, who admits thrill seekers into a fun-fair maze and then, after a period of twenty minutes or so, observes their leaving, subsequently wishes to determine which path they have taken. Of course, he cannot discern exactly where they have wandered without having peered inside; all he knows is that they have entered the labyrinth through the in door, exited through the out door, and taken one of thousands of possible routes through the maze. If he is very clever, he might map out their wave functions indicating the probabilities of their having trod on each of the possible paths.

Normally, this quantum mechanical approach to measurement, called the *sum over histories* method, is reserved for situations involving small particles. However, in the case of dual self-consistent solutions to wormhole time travel problems, such as in the billiards and salesman examples, it seems that this alternative technique is entirely appropriate.

Therefore, Thorne and his fellow wormhole designers have reached the conclusion that the best way of explaining the existence of multiple means of solving questions involving time travel is the use of quantum mechanical reasoning. The fact that one cannot say exactly which route the billiard ball or salesman takes indicates a fundamental difference between trajectory problems involving CTCs and tasks involving determination of paths taken in more simple geometries (without the possibility of loops). According to Thorne, in the former case quantum mechanics must be used, whereas in the latter a classical, pre-Heisenberg approach can be utilized. Traversable wormholes thus represent a rare large-scale application of a yet to be fully formulated quantum theory of gravitation.

In this quantum approach, the position and velocity of an object such as a billiard ball near a wormhole would no longer be quantities that could be exactly determined for all times. Instead, all the information about the ball would be contained in a wave function; this function would contain the probabilities for the ball to travel along each of the possible paths. In other words, it would be a sort of evolving probability distribution curve for the location of the billiard ball, indicating the chances for it to be at any given place with any given speed at any particular time. Of course, only self-consistent paths would be included in this curve—other routes would be ruled out by Thorne's principle of self-consistency.

So, in this formulation, as in the case of Schrödinger's puzzling cat, an object passing through a closed timelike curve would be considered to be in a mixed state, a limbo encompassing all of the possible paths. Until an experimental reading was taken, its true location would be a sort of haze spread out over all possibilities. Only if direct measurements of the object's position or momentum were made would its wave function collapse to one of the possibilities.

For example, in the salesman case, an electric eye could be set up to measure whether or not he deviated at all from following a straight path. If he were kidnapped by another version of himself, then the eye would measure a negative result. Otherwise, it would read positive, indicating that he hadn't strayed. Then, according to the standard interpretation of quantum mechanics, before the results were found the salesman could be said to be caught in a limbo between two possibilities: the one case being that he had walked absolutely straight; the other case, that he had been forced into the telephone booth wormhole by a version of himself that had already traveled through time (these are the zero loop and one loop paths). The salesman's wave function would be a juxtaposition of these two situations. Only when the electric eye results became known could the wave function be said to have collapsed into one of the two possible states.

*P*arallel Universes

The idea of collapsing wave functions and mixed states has become an integral part of the standard approach to quantum mechanics, known as the *Copenhagen interpretation*. However, even after over sixty years of successful continued use of this method, it remains steeped in controversy. The issue that many still find troubling is the idea that the observer should play such a significant part in the theory. As we have discussed, until an experimental measurement is taken, all objects, whether animate ones, such as salesmen or cats, or inanimate ones, such as billiard balls, are considered to be in fuzzy, mixed states of the quantities (position, momentum, and so forth) under consideration. Only after the investigator plays an active role, namely by taking a reading, does the wave function for this object collapse.

Because of this disturbing situation, many alternative models to the Copenhagen interpretation have been developed. Among the most intriguing, and highly controversial (far more so than the Copenhagen method), is an approach proposed by Hugh Everett in 1957, while he was completing his Ph.D. thesis requirements at Princeton University, and further developed by Bryce De Witt of the University of Texas. Usually referred to as either the *many-universes* or *many-worlds theory*, this novel way of interpreting quantum mechanics completely eliminates the role of the experimenter in influencing the behavior of the wave function.

In the many-universes approach, the wave function is treated as the most fundamental natural entity, rather than as simply a mathematical abstraction representing a distribution of probabilities. Then, the fact that this function is spread out over many possibilities simply means that these alternatives all actually exist. Finally, after a measurement is made, the universe branches and the experimenter finds himself or herself situated in one of the many copies. This process of bifurcation occurs again and again at each measurement of any physical quantity. In the words of De Witt: "Our universe must be viewed as constantly splitting into a stupendous number of branches. . . . Every quantum transition taking place on every star, in every galaxy, in every remote corner of the universe is splitting our local world into myriads of copies of itself. Here is schizophrenia with a vengeance!"

Even one's own body and consciousness must undergo billions and billions of splittings during one's lifetime, according to this radical approach. Each branching of the wave function of one of the body's constituent atoms results in a duplication of the self into versions situated on each of a myriad worlds.

Why aren't we aware that this world-splitting process is taking place? Even though, according to Everett and De Witt, we are constantly replicating ourselves, each of these "clones" is unaware of the others' existence. These replicas simply carry out their lives in alternate universes, with no way of being contacted nor of accessing the other worlds themselves. Like inmates serving time in solitary confinement at a mammoth prison, they have no way to find out what their comrades are doing, nor for them to find out what we're doing.

Moreover, according to John Wheeler, formerly a leading advocate of the many-universes theory but now a bit skeptical about whether such a radical approach is necessary, the universe itself splits as well. In his superspace scheme, there are billions and billions of versions of the cosmos, each evolving in a different manner. In some of these copies the galaxies are much larger than in ours; in others there are no galaxies at all; some versions collapse after only a few billion years; others last much longer; and so on.

How, then, can it be explained that we live in our specific universe, with its unique size and quantity of galaxies and its characteristic duration and behavior? To justify the distinctive nature of our particular world, many physicists and philosophers have proposed the *anthropic principle* as a means of singling out our own cosmos. This theory, combined with Wheeler's superspace proposal, envisions a collection of universes, each with its own special properties, evolving over time. Clearly then, amid the vast array of cosmic duplicates, there are at least a few universes that contain worlds able to sustain intelligent life. Only in those particular replicas would the cosmological conditions be right for beings like us to develop. This explains why our universe is as it is, in terms of its longevity and physical constants, for otherwise no conscious, intelligent beings would be around to record this information. In most of the other duplicate cosmos, no humans nor other self-aware organisms would exist—hence, no one could ever claim these hostile, alternate existences to be his own.

Although Wheeler and many of the other original proponents of the many-universes model have distanced themselves from its most far-reaching implications, this radical approach to quantum mechanics has had a strong influence on contemporary physics. The anthropic multiworlds approach to cosmology has won many prominent supporters, including Stephen Hawking. It's clear that the innovations encompassed by this theory will continue to play an important role in theoretical physics in the coming years.

Truly, the many-universes model represents a remarkably novel way of looking at time itself. Time, according to the worldview of Everett and De Witt, can no longer be conceived of as a single, flowing river. Rather, it must be viewed as a series of bifurcating streams, continuously branching out in all directions. Each of these forks represents one of the many possible changes that can take place in the cosmos.

This innovative view of time helps us to better understand many "paradoxes" in modern physics, such as the one concerning Schrödinger's cat (see Chapter 5). Recall that in the traditional view of this problem, this frazzled feline remains in a quantum limbo, suspended between life and death, until the experimenter opens its box and looks inside. Once this measurement is taken, the cat's wave function collapses to one of two possible values: alive or dead.

In the many-universes worldview, however, no such collapse occurs. Instead, the cat wave function and, in fact, the wave function representing the entire universe instantly divide into two parts. In one branch, the cat is definitely dead, in the other, unambiguously alive. Then, if the lab worker decided to lift up the cover of the container, he or she would be immediately propelled along one of the two possible forks. The experimenter would never know about the other possibility—would never feel the influence of the alternate version of the cosmos. The two worlds would never join up again and instead would eternally bifurcate into countless other realms.

Let's now return to the billiard ball problem and the related situation of the self-kidnapped salesman. It's apparent how we might apply the reasoning used in the cat paradox to resolving these wormhole time machine dilemmas. The idea of branching universes seems to lend itself nicely to resolving some of the perplexing situations generated by time travel, especially of the wormhole variety.

Indeed, the converse is also true: wormholes might provide considerable insight into the fundamental nature of quantum processes. Graham Collins of *Physics Today* has remarked that wormhole time travel might be a way of testing the validity of the many-worlds scheme: that by embarking on a trip into the past one might be able to discern whether or not wave function collapse takes place. At any rate, as the billiards and salesman examples show us, it's clear that there is a significant relationship between wormhole and multiworld notions.

Consider the billiards problem as translated into many-universes lingo. A billiard ball is projected into the space between two mouths of a wormhole. At that point, the universe branches into two parts. The first fork consists of a cosmos in which the ball travels along a completely straight path. On the other hand, in the second branch, the billiard ball travels through the wormhole after being bounced by another version of itself into one of the mouths. It then emerges out the other end, collides with itself again, and finally heads straight out away from the wormhole. Instead of one self-consistent universe, the cosmos has divided into two self-consistent realms in which there is only one possible solution per domain. The principle of self-consistency has been satisfied; hence, all paradoxes and double-solution dilemmas have been avoided.

The same logic can be applied equally well to the salesman puzzle. Our tale of the self-kidnapping salesman can be divided into two self-consistent stories, each corresponding to its own world. In one realm, Robert Preston, the salesman, continues directly along the path from his home to his office. In the other, Robert meets an earlier version of himself, is pushed into the phone booth by him, emerges from the bush, then keeps going. Each Robert continues through life, completely unaware that a bifurcation has taken place and that there is another Robert in another world. Hence, all contradictions have been successfully avoided.

It's clear that the many-universes theory serves to reduce many of the inconsistencies inherent in the notion of time travel via cosmic wormholes as well as to eliminate most of the problems associated with wave function collapse in quantum mechanics. Certainly this novel, unconventional view of reality yields considerable benefits in the realm of theoretical physics.

Yet, one might ask, is it worth the cost? Must we abandon the idea that our world is unique, and instead imagine that it is continually producing trillions of nearly identical copies, simply to help smooth over some of the perceived flaws in current models of cosmology and quantum mechanics? Perhaps there are other, less philosophically troubling, ways of ironing out these difficulties.

The multiworld model remains, to this day, a topic under considerable debate, with many physicists shifting camps from one side to another, much like a billiard ball batted between two mouths of a wormhole. Nevertheless, even if the many-universes model remains outside the realm of mainstream physics, its bold conceptions will most likely continue to capture the imagination as have few other models of the cosmos. Already, the notion that time is constantly branching has inspired numerous essays and fictional works, many scribed even before Everett presented his startling thesis. And naturally science fiction authors have led the way in producing innovative variations on this fascinating theme.

One of the first references to forking time was in a tale by David R. Daniels that appeared in 1934 in the science fiction magazine *Wonder Stories*. Here, Daniels uses the idea of branching universes to resolve a time travel paradox: by a well-placed switch to an alternate cosmos all contradictions in the story are averted. Each trip into the past, according to him, involves a forking of the cosmos along two paths—one containing "conventional" history, the other encompassing the altered version of reality.

In "The Legion of Time," by Jack Williamson, a short story that appeared four years later, the many-universes theory is anticipated in a portrayal of a world in which quantum mechanics guarantees a continuous splitting along branches of varied probability. Characters ponder the insignificance of death in an ever-branching cosmos, for even if a man is killed in a bus accident in one fork, he might still survive the crash in another.

Forking time has appeared in numerous other speculative tales. Considered by many critics the classic example of the branching universe genre, "The Man Who Folded Himself" by David Gerrold is, according to the science writer John Gribbin, "the definitive tongue-in-cheek ultimate exploration of the possibilities of an infinite array of almost-but-not-quite-the-same alternate probability worlds." Many of Philip K. Dick's novels, such as *Valis* and *The Man in the High Castle*, similarly involve the idea of parallel universes. Even Mark Twain has discussed this intriguing concept in *The Mysterious Stranger*.

The 1960s television series *Star Trek* provided an ideal canopy for the imaginative musings of science fiction writers and featured several prominent references to branching time. In "City on the Edge of Forever," written by Harlan Ellison, an inadvertent journey to the past strands the series's stars on 1930s earth. By saving the life of a prominent pacifist, one of the characters launches a chain of events that eventually leads to the victory of the Nazis in World War II. As a result, the *Star Trek* crew is temporarily diverted onto another fork of the universe until the time-transforming deed is undone. "Bread and Circuses," on the other hand, depicts an alternative world, albeit in our own universe, in which an exact replica of the Roman empire manages to acquire modern technology. Finally, in "Mirror, Mirror" the characters discover near likenesses of themselves in a strange, looking-glass copy of the cosmos.

Branching world stories are not confined to science fiction. In "The Garden of Forking Paths," a highly original detective story penned by the noted Latin American author Jorge Luis Borges, the theme of the universe as a labyrinth of possibilities is successfully explored. One of the people portrayed in this tale speaks to another of a relative who is obsessed with the notion of bifurcating time:

> In contrast to Newton and Schopenhauer, [he] did not believe in a uniform, absolute time. He believed in an infinite series of times, in a growing, dizzying net of divergent, convergent and parallel times. This network of times which approached one another, forked, broke off, or were unaware of one another for centuries, embraces all possibilities of time. We do not exist in the majority of these times; in some you exist, and not I; in others I, not you; in others, both of us.

This concept of the universe as a maze of alternatives is reflected in the writings of the acclaimed Italian novelist Italo Calvino. In *Invisible Cities*, he writes of parallel versions of the same city, each with a slightly different characteristic feature. This theme is further explored in *The Castle of Crossed Destinies*, a work produced by him with the help of a deck of Tarot cards; Calvino randomly distributed them and used their symbolism to tell a story. By using this method to generate his narrative, he hoped to show how arbitrary the sequence of world events is and how easy it is to imagine other groupings of the same elements. Calvino concludes that history, as we know it, is entirely determined by chance:

> The world does not exist . . . there is not an all, given all at once: there is a finite number of elements whose

combinations are multiplied to billions of billions, and only a few of these find a form and a meaning and make their presence felt amid a meaningless, shapeless dust cloud; like the seventy-eight cards of the tarot deck in whose juxtapositions sequences of stories appear and are then immediately undone."

Does the many-worlds theory present an actual experimentally verifiable depiction of our own physical universe, or is it simply a remarkably shrewd bit of mathematical hocus pocus with no apparent bearing on reality? Perhaps the study of traversable wormholes will soon yield an answer to this fascinating question. Until then, or whenever this issue is resolved by some other method, it is almost certain that the debate on the existence of multiple universes will continue to produce an ever-forking stream of journal articles devoted to this curious topic.

*W*hither Wormholes?

For the cosmic wormhole team of Morris, Thorne, Yurtsever, Novikov, and company, constructing a new method of interstellar transport has turned out to be much like building a high-rise office tower. In the case of skyscraper construction, a well-defined procedure must be followed by any architectural firm. First, a use must be found for the building; clients must be sought out and their needs determined. Second, detailed blueprints must be drawn up, and the safety and comfort of the potential occupants must be ensured. Third, the correct building materials have to be located and transported to the construction site. Fourth, a strong foundation suitable to support an immense edifice must be laid. Finally, after all these steps, the rest of the skyscraper can be erected.

In the mid-1980s, Kip Thorne, after consulting Carl Sagan, began a detailed determination of the need for traversable cosmic wormholes. His conclusion, reached along with Michael Morris, Ulvi Yurtsever, and Igor Novikov, was unmistakable: without some type of interstellar shortcut, rapid space travel to the distant planets would be virtually impossible. Black holes' connections, he reckoned, could never be rendered safe and stable enough for future use. Clearly, there was a strong need for sturdy, stable cosmic gateways. This was the "wormhole use assessment" phase: Stage One.

In 1988 and 1989, Thorne, Morris, and their fellow "architect" Matt Visser published a series of detailed blueprints for traversable wormholes. As in the case of competing plans for the same project, their constructs differed mainly in aesthetics, rather than in function. They carefully designed their creations to serve humanity in the best way possible, with comfort, speed, durability, and safety as their primary considerations. This was the "wormhole design" phase: Stage Two.

To locate and transport the raw materials needed for wormhole gateways was their next task. Exotic matter, of subzero mass, seemed to be the key to gateway construction. Yet it was most unclear how to find and haul this bizarre sort of material. In fact, none of the scientists was absolutely sure of its physical existence. Therefore, the next, "materials assembly" step, Stage Three, was never completed.

Nevertheless, the researchers proceeded to Stage Four, namely, laying the philosophical foundations for their important project. Much groundwork had to be accomplished before such an enormous undertaking could be completed. Only after this supporting theoretical structure was built could Stage Five, the actual construction of physical wormholes themselves, begin. Thus the last step of the wormhole-building project remains on hold until Stage Four (along with Three) is finally completed.

Bouncing around hypothetical cosmic billiard balls has played an essential role in this philosophical phase. Until this seeming paradox is fully understood, it will remain unclear whether wormhole theory rests on bedrock or quicksand. The entire science of gateway building will sink or stand depending on whether or not all of the implications of the existence of such constructions are physically valid. Thus, unless the self-interacting billiard ball puzzle and other time travel dilemmas are adequately resolved, all of the wormhole program's foundations are shaky.

Understandably, Thorne and his co-workers are eager to prove that their principle of self-consistency untangles the knotty paradoxes associated with closed timelike loops: the acausal structures endemic to wormholes of all sorts. Furthermore, they are eager to show how a full application of quantum mechanics to gravity would wholly explain how multiple, alternative self-consistent trajectories are possible for the same object passing close to the mouths of such gateways.

It seems that the controversial many-universes theory should play some role in all this. The concept of wormhole creation appears to open up the door to the possibility of parallel worlds. Whether these inaccessible, neighboring realms actually physically exist or simply abide in the netherworld of higher mathematics is a question that, it is hoped, will be satisfactorily resolved in coming years.

Once all of these philosophical issues and the problems associated with the use of exotic matter are finally settled, then the real task of assembling construction crews and actually building these interstellar passageways shall begin. Laborers willing to endure the harsh conditions of deep space will be recruited and sent out into the interstellar void to manipulate vast quantities of exotic material. Step by step, piece by piece, wormhole gateways will be fashioned according to their detailed blueprints, ensuring that these tunnels, once created, will enable future passengers to soar rapidly through the enormity of the cosmos in a way impossible for the original constructors of these shortcuts. Finally, these transgalactic connections will be finished and the new age of functional wormhole technology will begin.

Ultimately, wormhole construction may prove to be an arduous task, requiring the skills of hundreds of generations of "architects" and "builders," all following in the footsteps of the master-constructor Kip Thorne. Or it may, in fact, be a relatively simple feat, requiring, let's say, five hundred years or less—a few pages in the annals of human history. In either case it's abundantly clear that these monumental edifices, once completed, will enable humanity to break away at last from the confining environs of earth's once-isolated planetary village.

CHAPTER 9

SCENES FROM THE NEXT MILLENNIUM

Landing

A jarring noise breaks the silence of the wind-swept field. Rumbling and screaming, a jet touches its wheels on the long, straight runway, striking the tarmac with increasing sureness. Suddenly, the plane stops, and you disembark.

It's been twenty years since you left the town of your birth. Since that time it has awakened from a dormant, isolated desert community to become a thriving metropolis, a global center attracting visitors from far and wide. Walking from the airport to the periphery of the new city, you are stunned by the proliferation of high-rise buildings, their shiny exteriors illuminated by the setting sun in a dazzling display of color. Entering the community, you can barely recognize the once somber Main Street, now full of foreign merchants and their precious, exotic wares. Baskets of unusual handcrafted items from all over the world line the busy walkway, while the delicate scents of herbs and spices fill the twilight breeze. The sounds of exotic musical instruments mingle with the calls of the merchants in a melodious symphony, as you stroll down the once-silent high street.

Finally you reach the central square of the metropolis and stop for a while to reflect on the amazing transformation of your hometown from a sleepy isolated village to a bustling center of trade and industry. All of these changes were brought about by the existence of high-speed transport between your town and other communities and regions. Without the airplane, all would have remained as it was for generations: stagnant, unimaginative, and narrow. The ability to travel from one cultural enclave to another has meant the spreading of tradition and knowledge, the sharing of craftsmanship—in short, a reaching out to the world community in a mutually beneficial network of science and the arts. A once impassable frontier has been breached.

It is now nighttime. You gaze at a gleaming skyscraper in the distance and follow it upward with your eyes. Higher and higher it goes, pushing farther and farther out from its foundations. Your thoughts also soar higher and higher, following the shiny spire to dizzy new heights. Humanity is like the skyscraper, you think. It is always trying to reach out from its foundations and extend itself to new pinnacles of knowledge; always pushing ahead, always crossing new barriers. Where to go next, you wonder.

These are your thoughts as you scan the mighty tower, impressive against the background of a million glistening stars embedded in the tapestry of night.

*T*ours of the Universe

The tallest freestanding structure in the world today is the Canadian National (CN) Tower in Toronto. A lofty pillar of solid white concrete over 1,815 feet high, it stands as an impressive monument to humanity's age-old quest to reach the heavens. This giant edifice is topped by a spire and a bubble-shaped observation deck called, appropriately enough, the "space deck."

There is a long history of tower building as a means of extending the limits of man's vertical horizons. Traditionally the creation of large edifices has been viewed as a means of escaping the confines of the earth and reaching out to the distant celestial realm. These aspirations date back to Biblical times. It is written in Genesis that people dwelling in the land of Shinar tried (with disastrous results) to erect a "tower with its top in heaven" to symbolize their tribe's accomplishments. This Tower of Babel was seen by them as a way to unite and conquer the great divide between land and sky.

Visitors to the CN Tower have the choice of two opportunities to soar into the heavens. One option is to board a glass-enclosed elevator and take a high-speed ride up to the space deck for a breathtaking panoramic view of metropolitan Toronto. The other choice is located, strangely enough, within the bowels of the tower's base.

Buried within the core of the CN complex is a remarkable futuristic entertainment center called Tour of the Universe. Visitors there have an opportunity to imagine themselves as space explorers on a twenty-first-century pleasure cruise through the cosmos. Passengers for these simulated flights are greeted by friendly, colorfully garbed crew members who present them with intergalactic passports. To complete the illusion of passing through a spaceport, visitors then undergo a customs procedure which includes a laser-beam inoculation against extraterrestrial viruses such as "Martian dropsy" and "Ganymede rash," a flight briefing in which their itinerary is outlined, and a detailed clearance check. Finally, they board a state-of-the-art 727 flight simulator for a remarkably authentic trip through space, physically experiencing many of the jolting sensations of acceleration associated with rocket travel, while watching a specially produced film by Douglas Trumbull (who also designed the special effects for *2001: A Space Odyssey* and *Close Encounters of the Third Kind*).

In the imaginative drama that unfolds, the "Hermes Class IV shuttle" blasts off through the stem of the CN Tower, soars through deep space, and emerges in the vicinity of Jupiter. Of course, this takeoff isn't real; it's merely a highly sophisticated illusion. Visual and audio stimuli create an atmosphere of adventure and suspense as the passengers experience a near-collision with meteors as well as a brief interlude in "hyperspace." Finally, the "ship" lands and the travelers disembark.

Hundreds of thousands of tourists each year take the time to experience realistic spaceflight adventures such as the Toronto Tour of the Universe and other simulated interplanetary journeys in other cities. They vicariously experience the thrill of leaving planet earth behind and soaring into the unknown blackness of deep space. Clearly, there is a great unfulfilled need to explore the limitless cosmic frontiers— a desire that manifests itself in the unquenchable public interest in space stories, films, and flight simulators. Stargazers around the world are inspired each night by the wondrous magnitude of the visible universe, manifestly depicted in the thousands of points of light partially illuminating the darkened sky. Many long to fly upward toward this spectacle, to live the dream of spaceflight themselves.

The nighttime stellar canopy is indeed enticing—seemingly within reach but actually trillions of miles away. It taunts us to explore its nether reaches then recedes as we try to touch it. Only when humanity can experience firsthand the fiery powers of the ancient red suns, soar with grace among the outlying constellations, and taste the delicate fruits of distant planets, will this longing be satisfied.

*I*ntergalactic Switchboards

Someday rapid interstellar flight will no longer be a plot device for science fiction writers and amusement ride designers but an integral part of the human experience. Sooner or later, the technology will be developed to build gateways to other parts of the cosmos: traversable wormholes, perhaps of the sort envisioned by Thorne, Morris, Yurtsever, and Visser. These cosmic connections will allow human civilization to expand beyond the confines of the earth and to spread throughout the galaxy.

We stand at the dawn of the third millennium, a new and pivotal chapter of humankind's epic adventure. Barely one thousand years ago Leif Ericson and his band of Viking explorers were the first Europeans to land on the coast of North America. Almost five hundred years later, Columbus explored the American coast and prepared the way for the first European colonies and the Age of Exploration. These perilous missions opened up vast new frontiers for Western civilization and led to a great expansion of trade and industry. New forms of plant and animal life were found, new medicines produced, new cultures discovered. Are we ready yet to bridge the last frontier and to follow in the footsteps of our pioneering ancestors? Are we prepared to begin the second Age of Exploration, this time across the cosmic seas? Only the next millennium will yield the answers to these important questions.

Let us now try to picture what a new Age of Exploration would be like and imagine a future civilization built on rapid interstellar transport. The following is meant to be a speculative look at life in the year 2500:

Our account begins at a spaceport of the third millennium situated on earth and used for extraterrestrial voyages. Corridors in this massive building lead in all directions. Some are marked with the names of terrestrial locales; however, one has the designation "Earth Orbit Space Station." Passengers head through this latter gateway trailed by numerous pieces of cargo, each effortlessly carried on magnetic levitation devices. Once inside this portal, rapid elevators carry these travelers to the space transport level of the airport. After a brief health check, they pass down a dimly lit walkway to the entrance of a space shuttle.

Once inside the space shuttle, travelers are presented with a few choices. Some of the passengers elect to spend the spaceflight awake in the comfortable front cabin. Others choose to experience the tranquil state of cryonic suspension and move to the back section of the shuttle. There they are placed in carefully monitored cryogenic devices. These units are portable and marked with the travelers' final destinations. Since most of the hibernating voyagers are planning to transfer to other craft several times during their flight, all of these units are automatically shunted to the new spaceships at the appropriate transfer points. Thus many of the passengers taking journeys of a few weeks or longer can elect to spend these weeks in restful sleep, unhampered by the burdens of flight transfers.

Those travelers who decide to remain awake then prepare themselves for lift-off. After a wait of a few minutes, the shuttle speeds down a long runway, then rises into the atmosphere. Once in the air, the nuclear fusion engines are turned on and the shuttle reaches the Earth Orbit Space Station in a matter of minutes.

The space station is a colossal structure, about a mile across, and it orbits the earth approximately once every twenty-four hours. It serves multiple purposes: the flat surface on one side of the station is a landing strip for space shuttles arriving from earth; another face of the structure is a lift-off point for interplanetary spaceships.

In addition to its function as a way station, the mammoth craft contains permanent facilities for commerce, recreation, industry, and science. For the health and comfort of those aboard, the space station has a regulated breathable atmosphere. Inside the station are dormitories, lounges, and entertainment sectors.

Since the vessel is a gravity-free environment, many of the activities aboard are designed for zero gravity conditions. For example, international amateur and professional sports teams engage in weightless competitive activities, such as three-dimensional soccer, in the expansive cubical gym. Nearby, in a hospice ward, hundreds of injured and disabled individuals experience rapid recovery under the ideal circumstances of zero weight.

In yet another section, weightless manufacturing takes place and flawless crystals of perfect symmetry are grown to specification. Under zero gravity conditions, it is possible to manufacture products that are impossible to produce on earth because of gravitational constraints. Similarly, crystals so delicate that they would collapse under their own weight on earth can be grown in space.

Genetic research is carried out in a neighboring room; in the absence of gravity, medicines can readily be created by assembling benign "viruses" from proteins and chains of DNA, also manufactured in space. These genetic intruders are custom-made to "invade" the body and cure specific diseases. Forming these highly specialized viruses would be an arduous task on earth because of the devastating effects of gravity on these delicate organisms; in space, however, the problems associated with gravity can be prevented and genetic material can be readily constructed.

Some of these activities occur in special vacuum chambers isolated from the rest of the station. The rest take place in the much larger, atmospherically controlled section; the environment of the entire vessel is regulated by a complex computer network. This system also controls the trajectories of the hundreds of arriving and departing spaceships.

One such departing space vehicle is being prepared on a lower deck. After it has been fueled and inspected and all hibernation units and cargo have been moved on board, passengers enter the stellar cruiser, The Hawking, for the first leg of their transgalactic journey. The hatches on the ship are then closed, and the vehicle carefully detaches itself from the space station.

Passengers prepare themselves for the first blasts of The Hawking's acceleration units. A short jolt and the ship is traveling at ten thousand miles an hour; another blast and this rate is tripled. Another series of explosions occurs and now the cruiser's velocity begins its climb to its ultimate speed of 75 percent the speed of light.

, The Hawking's ultimate destination lies far beyond the farthest reaches of the planets, over ten times as far away from the sun as Pluto. However, even though it is 100 billion miles away from earth, the object to which the ship is traveling is part of the solar system. The spaceship Hawking is heading toward a space station near an interstellar gateway: a man-made wormhole orbiting the sun in a distant ellipse.

It takes the ship slightly over a week to complete its mission. For the hibernating voyagers, this week passes extremely rapidly; for those who have chosen to remain alert, numerous diversions help pass the time. Most of the passengers enjoy the chance to relax and do nothing for a week; many of them are on vacation and relish the peace and quiet of deep space.

Finally The Hawking reaches its destination: the Wormhole Orbit Space Station (WOSS), orbiting a three-mile-wide wormhole. Although this wormhole is much, much smaller than the sun, it is about twice as massive. If it were closer to the heart of the solar system, its mass would create a universally devastating effect, setting all of the planetary motions off balance. However, because it is so far away, it has only a minor gravitational effect on the motion of the sun, earth, and other planets.

The Hawking docks with the WOSS in a careful computer-monitored procedure. Its passengers disembark, and the cargo and cryonic suspension units are transferred to a holding area for the next flight. The voyagers walk over to the waiting room and remain there until the next leg of their journey can begin.

The Wormhole Orbit Space Station serves multiple functions, all related to the cosmic gateway that it orbits. Its primary purpose is to house the Transgalactica, a specially designed shuttle craft that transports passengers through the wormhole. In addition, workers and computers on board the WOSS carefully monitor the wormhole entrance, making small adjustments periodically to ensure that it doesn't collapse. If required, robot cargo ships can be sent out to the wormhole's throat, carrying quantities of exotic matter in special storage units. In the event that a repair is needed, the robot ships quickly unload the exotic material at the appropriate points on the throat, before being ripped apart by the strong gravitational fields in the area. Hence part of the WOSS is designated as an emergency repair base, harboring the robot ships.

An observation deck on one side of the WOSS allows visitors to look through special telescopes and gaze at the inner planets of the solar system from a vantage point 100 billion miles away. Thus the WOSS serves both pragmatic and recreational purposes.

At last the Transgalactica is ready for takeoff. All of the voyagers seeking to travel to other parts of the galaxy enter the ship and are placed in special safety restraints. Computers on board the ship set a course for the exact center of the wormhole face. The Transgalactica then blasts off, moving faster and faster, beginning its rapid descent into the heart of the cosmic gateway.

Within several hours, the ship has hit the wormhole surface, a safe part of the throat free of exotic matter. To the passengers on board, this entranceway looks like a dark rectangular slab, blocking out light from the nearby stars. They experience a brief jolt on entrance; soon, they are submerged in darkness: they are inside the throat. This experience lasts for just a few minutes, then suddenly the stars appear again in the sky. After a day-long flight out of the wormhole, the *Transgalactica* lands at another, larger space station, called Galaxy Central.

Galaxy Central is located in the heart of the Milky Way, a region where the sky is thick with stars. Surrounding it, though hundreds of millions of miles away, are thousands of wormhole entrances. These gateways, including the one through which the *Transgalactica* has just passed, lead to all parts of the galaxy. Thus Galaxy Central serves as a sort of transgalactic switchboard, coordinating flights to and from all regions of the cosmos.

Groups of passengers fill the corridors of this huge space complex with a strange mixture of languages and accents. In a typical example of the sort of interstellar travel that takes place these days, a family of Vegans scurries down a hall toward a Sirian-bound cruise vessel. Meanwhile, humans from the *Transgalactica* stop in the information center to inquire about available scenic flights to the Crab Nebula. A team of space scientists discuss the new sorts of minerals that have been found near Betelguese. Workers arriving from Cygnus X-1 mourn the accidental death of a crew member in an exotic matter mining expedition near a black hole. Another group of vacationing earthlings strolls to the *Transgalactica* dock on their way back to their home planet.

Suddenly there is a brief announcement: the *Transgalactica* is now boarding for the return flight to the solar system. A rush of travelers, of countless different nationalities and worlds of origin, heads toward the portal. Passengers are first screened for possible contaminants or space ailments and then allowed to board the ship.

The return flight of the *Transgalactica* is slightly different from the original voyage. Many of its occupants have recently returned from vacations, business trips, or trade expeditions in other parts of the galaxy. Some of them have spent months away from earth and would like to waste as little additional time as possible. Therefore, on the way back the *Transgalactica* makes a slight but useful detour: it heads toward a small wormhole specially designed for backward time travel.

By "slingshotting" through this portal and then heading back again through normal space, the clocks on the ship run backward for the equivalent of two weeks. Thus, the time lost on the flight to and from earth is recaptured on the passage through the time machine wormhole. At best, the interstellar flight will have taken literally no time at all.

Now, the *Transgalactica* begins the next leg of the return voyage. It moves rapidly through the original wormhole to the solar system, then emerges back near the WOSS. It docks at the space station and travelers transfer there for flights back to earth. Finally all of the voyagers set foot once again on terrestrial soil, carrying with them precious cargo and souvenirs from the various parts of the cosmos. Their exciting journeys have ended and they return home to rest.

*T*he Second Renaissance

Although this scenario is purely speculative, it does, I believe, provide some reasonable inkling as to what life in the next millennium might be like. Clearly, the emergence of traversable cosmic gateways would lead to a radical transformation of human society and might even acquaint us with the alien cultures that are bound to exist on the countless planets that most likely fill our galaxy. Moreover, the possibility of a new Age of Exploration gives us hope for a second Renaissance in art, music, literature, and science. The first Renaissance occurred in part because of the opening of new trade routes, particularly from Europe to the Orient and to the Americas. It seems fair to assume that the establishment of new interstellar passages would likely bring about a similar rebirth of cultural growth.

Even if no other societies are discovered among the distant stars, a new age would nonetheless herald vast innovations in geology, biology, medicine, and technology. If, however, alien races are encountered, who knows what variety of extraordinary changes might come about? Contacting advanced extraterrestrial civilizations might be the start of a process of colossal growth and innovation, rapid accumulation of new knowledge, and significant spiritual and ethical discovery as well.

The second law of thermodynamics suggests that the universe is heading toward an irreversible heat death and that eventually all stars will collapse to either charred fragments, dilute interstellar gases, neutron stars, or black holes. It is also possible that cosmic demise will someday come in the form of a cataclysmic big crunch. Dreary prospects all.

On the other hand, it is also possible that the centuries to come will bring about an era of unbridled creativity and innovation. Cosmic gateways could enable humankind to leave its terrestrial birthplace and soar freely among the heavens. And this would help to perpetuate the human race for eons to come, even if the earth itself were someday destroyed. Perhaps, in this manner, humanity would even be able to postpone its ultimate demise until time itself drew to a close.

The new era of traversable cosmic wormholes can be one of renewed strength and vitality. It is likely that the discovery of neutron stars, black holes, white holes, and quasars will lead to new and powerful sources of energy. It is even possible that time machines will be created from specially designed wormholes, enabling humanity to escape its temporal straitjacket and travel into its past or future, unlocking the secrets of both its own origins and destiny.

Cosmic wormholes, along with black holes, white holes, pulsars, and quasars, are some of the most enigmatic creations of modern scientific theory. They present extreme and baffling contrasts: unlimited production and total devastation, impenetrable barriers and traversable shortcuts, links to the distant past and visions of the end of time. The possible existence of these dents and tunnels in the spatial fabric forces us to confront our deepest fears and our wildest hopes. Only time, of course, will tell which destiny is ours: fate never reveals her secrets willingly, not even to physicists.

RELATED READING

Relativity

Bergmann, Peter G. *The Riddle of Gravitation*. New York: Charles Scribner's Sons, 1987.

Bohm, David. *The Special Theory of Relativity*. New York: Benjamin, 1965.

Chaisson, Eric. *Relatively Speaking*. New York: W. W. Norton and Co., 1988.

Einstein, Albert. *Relativity: The Special and General Theory*. New York: Crown Publishers, 1961.

Epstein, Lewis Carroll. *Relativity Visualized*. San Francisco: Insight Press, 1988.

Gribbin, John. *Spacewarps*. New York: Delacorte, 1983.

Misner, Charles, Kip Thorne, and John Wheeler. *Gravitation*. San Francisco: W. H. Freeman, 1973.

Pais, Abraham. *Subtle Is the Lord: The Science and Life of Albert Einstein*. New York: Oxford University Press, 1982.

Russell, Bertrand. *The ABC of Relativity*. London: Allen and Unwin, 1926.

Tauber, Gerald E. *Relativity: From Einstein to Black Holes.* New York: Venture, 1988.

Cosmology and the Early Universe

Barrow, John, and Joseph Silk. *The Left Hand of Creation: The Origin and Evolution of the Expanding Universe.* New York: Basic Books, 1983.

Ferris, Timothy. *Coming of Age in the Milky Way.* New York: William Morrow and Co., 1988.

Gribbin, John, and Martin Rees. *Cosmic Coincidences: Dark Matter, Mankind and Anthropic Cosmology.* New York: Bantam Books, 1989.

Harrison, Edward. *Cosmology: The Science of the Universe.* New York: Cambridge University Press, 1981.

Hawking, Stephen. *A Brief History of Time: From the Big Bang to Black Holes.* New York: Bantam Books, 1988.

Jastrow, Robert. *Red Giants and White Dwarfs.* New York: W. W. Norton and Co., 1979.

Sagan, Carl. *The Cosmic Connection.* Garden City, N.Y.: Anchor Press, 1973.

————. *Cosmos.* New York: Ballantine Books, 1980.

Silk, Joseph. *The Big Bang: The Creation and Evolution of the Universe.* San Francisco: W. H. Freeman, 1980.

Tucker, Wallace, and Kevin Tucker. *The Dark Matter.* New York: William Morrow and Co., 1988.

Weinberg, Stephen. *The First Three Minutes: A Modern View of the Origin of the Universe.* New York: Basic Books, 1977.

Space Travel and the Search for Extraterrestrial Life

Ryan, Peter, and L. Pesek. *Solar System.* New York: Viking, 1979.

Nicholson, Ian. *The Road to the Stars.* New York: William Morrow and Co., 1978.

Shklovskii, I. S., and Carl Sagan. *Intelligent Life in the Universe.* New York: Dell, 1967.

Time Travel and Parallel Universes

De Witt, Bryce, and Neill Graham, eds. *The Many-Worlds Interpretation of Quantum Mechanics.* Princeton, N.J.: Princeton University Press, 1973.

Gardner, Martin. *Time Travel and Other Mathematical Bewilderments.* New York: W. H. Freeman, 1988.

Gribbin, John. *In Search of Schrödinger's Cat.* New York: Bantam Books, 1984.

————. *Timewarps.* New York: Dell Publishing Co., 1979.

Halpern, Paul. *Time Journeys: A Search for Cosmic Destiny and Meaning.* New York: McGraw-Hill, 1990.

Herbert, Nick. *Faster than Light: Superluminal Loopholes in Physics.* New York: New American Library, 1989.

Nicholls, Peter. "Time Travel and Other Universes." In *The Science in Science Fiction,* edited by Peter Nicholls. New York: Alfred A. Knopf, 1983.

Wolf, Fred Alan. *Parallel Universes: The Search for Other Worlds.* New York: Touchstone Books, 1988.

Black Holes

Asimov, Isaac. *The Collapsing Universe*. New York: Walker and Co., 1977.

Berry, Adrian. *The Iron Sun*. New York: E. P. Dutton, 1977.

Chandrasekar, S. *The Mathematical Theory of Black Holes*. New York: Oxford University Press, 1983.

Greenstein, George. *Frozen Star*. New York: Freundlich Books, 1983.

Gribbin, John. *White Holes: Cosmic Gushers in the Universe*. New York: Delacorte, 1977.

Novikov, Igor. *Black Holes and the Universe*. New York: Cambridge University Press, 1990.

Shapiro, Stuart, and Saul Teukolsky. *Black Holes, White Dwarfs and Neutron Stars*. New York: John Wiley and Sons, 1983.

Shipman, Harry. *Black Holes, Quasars and the Universe*. Boston: Houghton Mifflin Co., 1980.

Sullivan, Walter. *Black Holes: The Edge of Space, The End of Time*. New York: Doubleday, 1979.

Taylor, John. *Black Holes: The End of the Universe?* New York: Avon Books, 1973.

Science Fiction

Asimov, Isaac. *The End of Eternity*. Greenwich, Conn.: Fawcett Crest Books, 1972.

Bellamy, Edward. *Looking Backward*. New York: Magnum Books, 1968.

Benford, Gregory, and Martin Greenberg, eds. *Hitler Victorious*. New York: Berkley Books, 1987.

Bloch, Robert. "That Hell-Bound Train." In *The Best of Robert Bloch*, edited by Lester del Rey. New York: Ballantine Books, 1977.

Borges, Jorge Luis. "The Garden of Forking Paths." In *Labyrinths*. New York: New Directions Publishers, 1962.

Boulle, Pierre. *Planet of the Apes*. New York: Signet Books, 1964.

Bradbury, Ray. "A Sound of Thunder." In *R Is for Rocket*. New York: Bantam Books, 1969.

———. "The Toynbee Convector." In *The Toynbee Convector*. New York: Bantam Books, 1989.

Brown, Fredric. "Experiment." In *The Best of Fredric Brown*, edited by Robert Bloch. New York: Ballantine Books, 1977.

Calvino, Italo. *The Castle of Crossed Destinies*. New York: Harcourt Brace Jovanovich, 1973.

Clarke, Arthur C. *2001: A Space Odyssey*. New York: Signet Books, 1968.

Dahl, Roald. *Charlie and the Great Glass Elevator*. New York: Alfred A. Knopf, 1972.

Dick, Philip K. *The Man in the High Castle*. New York: Ace Books, 1988.

Heinlein, Robert. "All You Zombies." In *Approaches to Science Fiction*, edited by Donald L. Lawler. Boston: Houghton Mifflin Co., 1978.

Irving, Washington. *Rip Van Winkle*. New York: William Morrow and Co., 1987.

Leiber, Fritz. *The Big Time*. New York: Ace Books, 1961.

———. "Try and Change the Past." In *Trips in Time*, edited by Robert Silverberg. New York: Thomas Nelson, 1977.

Moorcock, Michael. *Behold the Man*. New York: Carroll and Graf, 1987.

Moore, Ward. *Bring the Jubilee*. New York: Bart Books, 1988.

Sagan, Carl. *Contact.* New York: Simon and Schuster, 1985.

Twain, Mark. *A Connecticut Yankee in King Arthur's Court.* New York: P. F. Collier and Son, 1918.

Watson, Ian. *The Very Slow Time Machine.* New York: Ace Books, 1979.

Wells, Herbert G. "The Time Machine." In *Three Prophetic Novels.* Selected by E. F. Bleiber. New York: Dover, 1960.

Williamson, Jack. *The Legion of Time.* New York: Bluejay Books, 1985.

*T*echnical Articles

Benford, G. A., D. L. Book, and W. A. Newcomb. "The Tachyonic Antitelephone." *Physical Review* 2D (1970): 263.

Clarke, C. J. S. "Opening a Can of Wormholes." *Nature* 348 (1990): 287.

De Witt, Bryce. "Quantum Mechanics and Physical Reality." *Physics Today* 23 (1970): 4.

Eardley, Douglas M. "Death of White Holes in the Early Universe." *Physical Review Letters* 33 (1974): 442.

Friedman, John, Michael S. Morris, Igor D. Novikov, Fernando Echeverria, Gunnar Klinkhammer, Kip S. Thorne, and Ulvi Yurtsever. "Cauchy Problem in Spacetimes with Closed Timelike Curves." *Physical Review* 42D (1990): 1915.

Frolov, Valery P., and Igor D. Novikov. "Physical Effects in Wormholes and Time Machines." *Physical Review* 42D (1990): 1057.

Gödel, Kurt. "An Example of a New Type of Cosmological Solutions of Einstein's Field Equations of Gravitation." *Reviews of Modern Physics* 21 (1949): 417.

Gott, J. Richard III. "Closed Timelike Curves Produced by Pairs of Moving Cosmic Strings: Exact Solutions." *Physical Review Letters* 66 (1991): 1126.

Kerr, Roy P., and A. Schild. "A New Class of Vacuum Solutions of the Einstein Field Equations." In *Proceedings of the Galileo Galilei Centenary Meeting on General Relativity, Problems of Energy and Gravitational Waves*, edited by G. Barbera. Florence: Comitato Nazionale per la Manifestazione Celebrative (1965), 222.

Lightman, Alan. "Still Wanted: Black Holes." *Discover* (March 1990): 26.

Metzenthen, William E. "Appearance of Distant Objects to an Observer in a Charged Black Hole Spacetime." *Physical Review* 42D (1990): 1105.

Michell, John. *Philosophical Transactions of the Royal Society of London* 74 (1784): 35.

Morris, Michael S., and Kip S. Thorne. "Wormholes in Spacetime and Their Use for Interstellar Travel: A Tool for Teaching General Relativity." *American Journal of Physics* 56 (1988): 395.

———, and Ulvi Yurtsever. "Wormholes, Time Machines and the Weak Energy Condition." *Physical Review Letters* 61 (1988): 1446.

Redmount, Ian. "Wormholes, Time Travel and Quantum Gravity." *New Scientist* (April 1990): 57.

Tipler, Frank. "Rotating Cylinders and the Possibility of Global Causality Violation." *Physical Review* 9D (1974): 2203.

Unruh, William G. "Notes on Black Hole Evaporation." *Physical Review* 14D (1976): 870.

———, and Robert M. Wald. "Entropy Bounds, Acceleration Radiation and the Generalized Second Law." *Physical Review* 27D (1983): 2271.

Visser, Matt. "Traversable Wormholes from Surgically Modified Schwarzschild Spacetimes." *Nuclear Physics* B328 (1989): 203.

———. "Traversable Wormholes: Some Simple Examples." *Physical Review* 39D (1989): 3182.

———. "Wormholes, Baby Universes and Causality." *Physical Review* 41D (1990): 1116.

INDEX

MENTOR

THE MYSTERIES OF SCIENCE

☐ **UNDERSTANDING PHYSICS: The Electron, Proton, and Neutron by Isaac Asimov.** This is a brilliant picture of the men and ideas that have given our world the laser beam and the H-bomb, with their mixed legacy of wonder and fear. A fascinating study designed for students and laymen alike.
(626346—$5.95)

☐ **UNDERSTANDING PHYSICS: Light Magnetism, and Electricity by Isaac Asimov.** The great transition from Newtonian physics to the physics of today forms one of the most important chapters in the annals of scientific progress. Climaxing with Planck's and Einstein's landmark contributions, this immense expansion of knowledge is examined and explained. (626354—$4.95)

☐ **UNDERSTANDING PHYSICS: Motion, Sound, and Heat by Isaac Asimov.** Aristotle, Galileo, Newton—the drama of their discoveries, and the far-reaching ramifications of those epochal breakthroughs are placed in focus by a writer unexcelled in rendering science intelligible for the nonexpert. In this volume, the author charts the vital link between the scientific past and the scientific present, and opens the path to understanding a branch of knowledge supremely important in today's world. (626621—$5.99)

Prices slightly higher in Canada.

There's an epidemic with 27 million victims. And no visible symptoms.

It's an epidemic of people who can't read.

Believe it *or* not, 27 million Americans are functionally illiterate, about one adult in five.

The solution to this problem is you... when you join the fight against illiteracy. So call the Coalition for Literacy at toll-free **1-800-228-8813** and volunteer.

Volunteer Against Illiteracy. The only degree you need is a degree of caring.